Praise for
Reader, Come Home

"The author of *Proust and the Squid* returns to the subject of technology's effect on our brains and our reading habits. This is an even more direct plea and a lament for what we are losing, as Wolf brings in new research on the reading brain and examines how the digital realm has degraded her own concentration and focus."

—*New York Times* New & Noteworthy

"Wolf offers a persuasive catalog of the cognitive and social good created by deep reading. . . . She's right that digital media doesn't automatically doom deep reading, and can even enhance it. She's also correct that we have a lot to lose if we don't pay attention to what we're doing with technology and what it's doing to us."

—*Washington Post*

"[A] gentle manifesto. . . . [Wolf] affirms and celebrates the power of reading for the formation of our moral imaginations, and a lifetime of bookish devotion bubbles to the surface of her lovely prose in allusion and quotation."

—*Washington Free Beacon*

"Maryanne Wolf has done it again. She has written another seminal book destined to become a dog-eared, well-thumbed, often-referenced treasure on your bookshelf. . . . *Reader, Come Home* conveys a cautionary message, but it also will rekindle your heart and help illuminate promising paths ahead."

—International Dyslexia Association

"The digital age is effectively reshaping the reading circuits in our brains, argues Ms. Wolf. But there's hope: Sustained, close reading is vital to redeveloping attention and maintaining critical thinking, empathy, and myriad other skills in danger of extinction. Luckily, her book isn't difficult to pay attention to."

—*Wall Street Journal*

"Wolf is a lovely prose writer who draws not only on research but also on a broad range of literary references, historical examples, and personal anecdotes. The strongest parts of *Reader, Come Home* are her moving accounts of why reading matters, and her deeply detailed exploration of how the reading brain is being changed by screens. . . . Wolf makes a strong case for what we lose when we lose reading."

—*San Francisco Chronicle*

"This rich study by cognitive scientist Maryanne Wolf tackles an urgent question: How do digital devices affect the reading brain? Wolf explores the 'cognitive strata below the surface of words,' the demotivation of children saturated in on-screen stimulation, and the power of 'deep reading' and challenging texts in building nous and ethical responses such as empathy. . . . An antidote for today's critical-thinking deficit."

—*Nature*

"Timely and important. . . . If you love reading and the ways it has enriched your life and our world, *Reader, Come Home* is essential, arriving at a crucial juncture in history."

—*BookPage*

"Wolf wields her pen with equal parts wisdom and wonder. The result is a joy to read and reread, a love letter to literature, literacy, and progress."

—*Shelf Awareness*

"[*Reader, Come Home*] is necessary for these times, as . . . long-form writing is in crisis, and information is digested in small bits more than ever. With thought leaders like Wolf lending her thoughts to how we can preserve reading in its current form, we can have a brighter future. With *Reader, Come Home*, Wolf can now claim to have two groundbreaking works."

—*Brooklyn Digest*

"In this profound and well-researched study of our changing reading patterns, Wolf presents lucid arguments for teaching our brain to become all-embracing in the age of electronic technology. If you call yourself a reader and want to keep on being one, this extraordinary book is for you."

—Alberto Manguel, author of *A History of Reading*

"A love song to the written word, a brilliant introduction to the science of the reading brain, and a powerful call to action. With each page, Wolf shows us why we must preserve deep reading for ourselves and sow desire for it within our kids. Otherwise we risk losing the critical benefits for humanity that come with reading deeply to understand our world."

—Lisa Guernsey, coauthor of *Tap, Click, Read: Growing Readers in a World of Screens*

"Scholar, storyteller, and humanist, Wolf brings her laser sharp eye to the science of reading in a seminal book about what it means to be literate in our digital and global age. Informed by a review of research from neuroscience to Socratic philosophy, and wittily crafted with true affection for her audience, *Reader, Come Home* charts a compelling case for a new approach to lifelong literacy that could truly affect the course of human history."

—Michael H. Levine, coauthor of *Tap, Click, Read: Growing Readers in a World of Screens*

"Our best research tells us that deep reading is an essential skill for the development of intellectual, social, and emotional intelligence in today's children. In our increasingly digital world—where many children spend more time on social media and gaming than just about any other activity—do children have any hope of becoming deep readers? In her must-read *Reader, Come Home*, a game-changer for parents and educators, Maryanne Wolf teaches us about the complex workings of the brain and shows us when—and when not—to use technology."

—Catherine Steiner-Adair, author of *The Big Disconnect*

"This is a book for all of us who love reading and fear that what we love most about it seems to slip away in the distractions and interruptions of the digital world. Here we are challenged to take the steps to ensure that what we cherish most about reading—the experience of reading deeply—is passed on to new generations. Wolf is sober, realistic, and hopeful, an impressive trifecta. Her core message: We can't take reading too seriously. And for us, today, how seriously we take it, will mark the measure of our lives."

—Sherry Turkle, author of *Reclaiming Conversation: The Power of Talk in a Digital Age*

"Wolf stays firmly grounded in reality when presenting suggestions . . . for how to teach young children to be competent, curious, and contemplative in a world awash in digital stimulus. [*Reader, Come Home*] is a clarion call for parents, educators, and technology developers to work to retain the benefits of reading independent of digital media."

—*Publishers Weekly* (starred review)

"Wolf (Tufts, *Proust and the Squid*) provides a mix of re-assurance and caution in this latest look at how we read today. . . . A hopeful look at the future of reading that will resonate with those who worry that we are losing our ability to think in the digital age. Accessible to general readers and experts alike."

—*Library Journal* (starred review)

"Wolf's writing . . . is buoyed by encyclopedic knowledge of cognitive science and of literature, laced with insight, and infused with a rare mix of science and artistry."

—*EducationNEXT*

"In her book *Reader, Come Home*, Maryanne Wolf has adopted the discontinuity perspective to explore the changes in the process of reading with screens. Wolf issues a clarion call to remain vigilant about the deep reading processes that readers of all ages need to develop and maintain in the digital era."

—*Journal of Children and Media*

"Maryanne Wolf's *Reader, Come Home* returns after ten years to map a cognitive landscape that was only beginning to take shape in her earlier book, *Proust and the Squid*. In *Reader, Come Home*, Wolf is looking to understand how our brains might be adapting to a new type of reading, and the implications for individuals and societies. The book is a combination of engaging synthesis of neuroscience and educational research, with reflection on literature and literary reading."

—*The Scholarly Kitchen*

"Wolf is a serious scholar genuinely trying to make the world a better place. *Reader, Come Home* is full of sound . . . advice for parents."

—*Slate Book Review*

"The book is a rewarding read, not only because of the ideas Wolf presents us with but also because of her warm writing style and rich allusion to literary and philosophical thinkers, infused with such a breadth of authors that only a true lover of reading could have written this book. It is a necessary volume for everyone who wants to understand the current state of reading in America."

—*Stanford Daily*

"I've just finished reading this extraordinary new book . . . essential reading for anyone who has the privilege of introducing young people to the wonders of language, and especially those who work with children under the age of ten."

—*Englewood Review of Books*

"*Reader, Come Home* provides us with intimate details of brain function, vision, language, and neuroplasticity. Wolf has endeavored to make something extremely complicated more accessible and for the most part she succeeds. The book is written as a series of letters to you, the reader. If you are a parent, it will probably be the most important book you read this year."

—*Bookshelf*

"How often do you read in a deep and sustained way fully immersed, even transformed, by entering another person's world? Maryanne Wolf cautions that the way our engagement with digital technologies alters our reading and cognitive processes could cause our empathic, critical thinking, and reflective abilities to atrophy. This in turn could undermine our democratic, civil society."

—Learning & the Brain

"In this epistolary book, Wolf draws on neuroscience, psychology, education, philosophy, physics, physiology, and literature to examine the differences between reading physical books and reading digitally. . . . An accessible, well-researched analysis of the impact of literacy."

—*Kirkus Reviews*

Praise for
Reader, Come Home
from India

"In her new book, Wolf . . . frames our growing incapacity for deep reading. She . . . explains how our ability to be 'good readers' is intimately connected to our ability to reflect, weigh the credibility of information that we are bombarded with across platforms, form our own opinions, and ultimately strengthen democracy."

—*The Hindu*

"Wolf raises a clarion call for us to mend our ways before our digital forays colonize our minds completely."

—*Deccan Herald*

Praise for
Reader, Come Home
from Italy

"Maryanne Wolf goes to the heart of the problem: reading is a political act and the speed of information can decrease our critical thought."
—*Corriere della Sera* (Alessandro D'Avenia)

"The heart of this book brings us to our own 'deep reading' processes—the ability to enter into the text, to feel that we are part of it."
—*Il Sole 24 Ore*

"Neuroscience-based advice to parents of digital natives: the latest book by Maryanne Wolf explains how to maintain focus and navigate a constant bombardment of information."
—*La Repubblica*

"Are we able to truly read any longer? This is the question that Maryanne Wolf asks herself and our world."
—*Anderse*

"This last beautiful book of Maryanne Wolf both suggests that we protect children from screen dependency and also that we . . . need to give back the joy of the reading experience to our children!"
—*Corriere della Sera* (Carlo Ossola)

READER, COME HOME

ALSO BY MARYANNE WOLF

Proust and the Squid:
The Story and Science of the Reading Brain

Tales of Literacy for the 21st Century:
The Literary Agenda

READER,

COME

HOME

The Reading Brain in a Digital World

MARYANNE

WOLF

Illustrated by Catherine Stoodley

HARPER

NEW YORK · LONDON · TORONTO · SYDNEY

HARPER

FIRST HARPER PAPERBACKS EDITION PUBLISHED 2019.

Designed by Fritz Metsch

Library of Congress Cataloging-in-Publication Data has been applied for.

ISBN 978-0-06-238877-3 (pbk.)

22 23 24 25 26 LBC 13 12 11 10 9

To my mother, my best friend,
Mary Elizabeth Beckman Wolf
(June 26, 1920–December 5, 2014)

If we could modify the structure and wiring of the brain, that would be a fundamental game changer in terms of who we are, what we decide, what we think. . . . We are in a different phase of evolution; the future of life is now in our hands. It is no longer just natural evolution, but human-driven evolution.

—Juan Enriquez and Steve Gullans

Reading is an act of contemplation . . . an act of resistance in a landscape of distraction . . . it returns us to *a reckoning with time*.

—David Ulin

CONTENTS

CONTENTS

READER, COME HOME

Letter One

READING, THE CANARY IN THE MIND

Fielding calls out to you every few paragraphs
as if to make sure you have not closed the book,
and now I am summoning you up again,
attentive ghost, dark silent figure standing
in the doorway of these words.
—Billy Collins [my italics]

Dear Reader,

You stand at the doorway of my words; together we stand at the threshold of galactic changes over the next few generations. These letters are my invitation to consider an improbable set of facts about reading and the reading brain, whose implications will lead to significant cognitive changes in you, the next generation, and possibly our species. My letters are also an invitation to look at other changes, more subtle ones, and consider whether you have moved, unaware, away from the home that reading once was for you. For most of us, these changes have begun.

Let's begin with a deceptively simple fact that has inspired my work on the reading brain over the last decade and move from there: *human beings were never born to read.* The acquisition of literacy is one of the most important epigenetic achievements of *Homo sapiens*. To our knowledge, no other species ever acquired it. The act of learning to read added an entirely new circuit to our hominid brain's repertoire. The long developmental process of learning to read deeply and well changed the very structure of that circuit's connections, which rewired the brain, which transformed the nature of human thought.

What we read, how we read, and why we read change how we think, changes that are continuing now at a faster pace. In a span of only six millennia reading became the transformative catalyst for intellectual development within individuals and within literate cultures. The quality of our reading is not only an index of the quality of our thought, it is our best-known path to developing whole new pathways in the cerebral evolution of our species. There is much at stake in the development of the reading brain and in the quickening changes that now characterize its current, evolving iterations.

You need only examine yourself. Perhaps you have already noticed how the quality of your attention has changed the more you read on screens and digital devices. Perhaps you have felt a pang of something subtle that is missing when you seek to immerse yourself in a once favorite book. Like a phantom limb, you remember who you were as a reader, but cannot summon that "attentive ghost" with the joy you once felt in being transported somewhere outside the self to that interior space. It is more difficult still with children, whose attention is continuously distracted and flooded by stimuli that will never be consolidated in their reservoirs of knowledge. This means that the very basis of their capacity to draw analogies and inferences when they read will be less and less developed. Young reading brains are evolving without a ripple of concern by most people, even though more and more of our youths are not reading other than what is required and often not even that: "tl; dr" (too long; didn't read).

In our almost complete transition to a digital culture we are changing in ways we never realized would be the unintended collateral consequences of the greatest explosion of creativity, invention, and discovery in our history. As I chronicle in these letters, there is as much reason for excitement as caution if we turn our attention to the specific changes in the evolving reading brain that are happening now and may happen in different ways in a few short years. This is because the transition from a literacy-based culture to a digital one differs radically from previous transitions from one form of communication to another. Unlike in the past, we possess both the science and the technology to identify potential changes in how we read—and thus how we think—before such changes are fully entrenched in the population and accepted without our comprehension of the consequences.

The building of this knowledge can provide the theoretical basis for changing technology to redress its own weaknesses, whether in more refined digital modes of reading or the creation of alternative, developmentally hybrid approaches to acquiring it. What we can learn, therefore, about the impact of different forms of reading on cognition and culture has profound implications for the next reading brains. Thus equipped, we will have the capacity to help shape the changing reading circuits in our children and our children's children in wiser and better-informed ways.

I invite you into my collected thoughts on reading and the evolving reading brain as I would a friend at my door—with equal parts anticipation and delight at our exchanges about what reading means, beginning with the story of how reading became so important to me. To be sure, when I was a child learning to read, I did not think about reading. Like Alice, I simply jumped down reading's hole into Wonderland and disappeared for most of my childhood.

When I was a young woman, I did not think about reading.
I simply became Elizabeth Bennet, Dorothea Brooke, and
Isabel Archer at every opportunity. Sometimes I became
men like Alyosha Karamazov, Hans Castorp, and Holden
Caulfield. But always I was lifted to places very far from
the little town of Eldorado, Illinois, and always I burned
with emotions I could never otherwise have imagined.

Even when I was a graduate student of literature, I did
not think very much about reading. Rather, I pored over
every word, every encrypted meaning in the *Duino Elegies*
by Rilke and novels by George Eliot and John Steinbeck,
and felt myself bursting with sharpened perceptions of the
world and anxious to fulfill my responsibilities within it.

I failed my first round at the latter miserably and memorably. With all the enthusiasm a young, flimsily prepared
teacher can have, I began a Peace Corps–like stint in rural
Hawaii along with a small and wonderful group of fellow
would-be teachers. There I stood daily before twenty-four
unutterably beautiful children. They looked at me with
complete confidence, and we looked at each other with total, reciprocated affection. For a while those children and
I were oblivious to the fact that I could change the circumstances of their life trajectories if I could help them become
literate, unlike many in their families. Then, only then, did
I begin to think seriously about what reading means. It
changed the direction of my life.

With sudden and complete clarity I saw what would happen if those children could not learn the seemingly simple
act of passage into a culture based on literacy. They would
never fall down a hole and experience the exquisite joys of
immersion in the reading life. They would never discover
Dinotopia, Hogwarts, Middle Earth, or Pemberley. They
would never wrestle through the night with ideas too large
to fit within their smaller worlds. They would never experi-

ence the great shift that moves from reading about characters like the Lightning Thief and Matilda to believing they could become heroes and heroines themselves. And most important of all, they might never experience the infinite possibilities within their own thoughts that emerge whole cloth from each fresh encounter with worlds outside their own. I realized in a whiplash burst that those children, all mine for one year, might never reach their full potential as human beings if they never learned to read.

From that moment on, I began in earnest to think about reading's capacity to change the course of an individual life. What I hadn't a clue about then was the deeply generative nature of written language and what it means—literally and physiologically—for generating new thoughts, not only for a child but for our society. I also had no glimpse of the extraordinary cerebral complexity that reading involves and how the act of reading embodies as no other function the brain's semi-miraculous ability to go beyond its original, genetically programmed capacities such as vision and language. That would come later, as it will in these letters. I revised my entire life plan, and moved from the love of written words to the science beneath them. I set out to understand how human beings acquire written words and use written language to great advantage for their own intellectual development and that of future generations.

I never looked back. Decades have passed since I taught the children of Waialua, now grown with children of their own. Because of them, I became a cognitive neuroscientist and a scholar of reading. More specifically, I conduct research on what the brain does when it reads and why some children and adults have greater difficulty learning how to read than others do. There are many reasons, from external causes such as children's impoverished environments to more biological reasons such as differences in the brain's organization of language in the grossly misunderstood phenomenon of dyslexia. But these are themes in other directions of my work and will make only cameo appearances in this book.

These letters are concerned with a different direction of my work on the reading brain: the intrinsic plasticity that underlies it, with the unexpected implications that affect us all. My first inklings of the high stakes involved in the reading circuit's plasticity began more than a decade ago when I launched forth on what I thought would be a relatively circumscribed task: a researcher's account of reading's contributions to human development in *Proust and the Squid: The Story and Science of the Reading Brain*. My original intention was to describe the great arc of literacy's development and provide a new conceptualization of dyslexia that would describe the cerebral riches that are often wasted when people do not understand individuals whose brains are organized for language in a different way.

But something unexpected happened as I wrote that book: reading itself changed. What I knew as a cognitive neuroscientist and developmental psychologist about the development of written language had begun to shift before my eyes and under my fingers and under everybody else's, too. For seven years I had studied the beginnings of Sumerian scripts and Greek alphabets and analyzed brain-imaging data with my own brain largely buried in research. When I finished, I lifted my head to look about me and felt as if I were Rip Van Winkle. In the seven years it had taken me to describe how the brain had learned to read over its close to six-thousand-year history, our entire literacy-based culture had begun its transformation into a very different, digitally based culture.

I was gobsmacked. I rewrote the first, historical chapters of my book to reflect the striking parallels between our present cultural shifts to a digital culture and the similar transition from the Greeks' oral culture to their extraordinary written one. That was comparatively simple, thanks to a critical tutorial given to me by a very generous classicist colleague, Steven Hirsh. It was anything but simple, however, to use research on the existing, expert reading brain to predict its next adaptation. And that is where I stopped in 2007. My self-appointed role as narrator of the research world's insights into reading's mind-changing capacities had moved out of my ken.

There was almost no research being conducted then on the formation of a digital reading brain. There were no significant studies about what was happening in the brains of children (or adults) as they learned to read while immersed in a digitally dominated medium six to seven hours a day (a figure that has since almost doubled for many of our youth). I knew how reading changes the brain and how the brain's plasticity enables it to be shaped by external factors such as a particular writing system (e.g., English versus Chinese). Unlike scholars in the past such as Walter Ong and Marshall McLuhan, I never focused on the influences of the medium (e.g., book versus screen) upon this malleable circuit's structure. By the end of writing *Proust and the Squid*, however, I changed. I became consumed with how the circuitry of the reading brain would be altered by the unique characteristics of the digital medium, particularly in the young.

The unnatural, cultural origin of literacy—the first deceptively simple fact about reading—means that young readers do not have a genetically based program for developing such circuits. Reading-brain circuits are shaped and developed by both natural and environmental factors, including the medium in which reading is acquired and developed. Each reading medium advantages certain cognitive processes over others. Translation: the young reader can either develop all the multiple deep-reading processes that are currently embodied in the fully elaborated, expert reading brain; or the novice reading brain can become "short-circuited" in its development; or it can acquire whole new networks in different circuits. There will be profound differences in how we read and how we think, depending on which processes dominate the formation of the young child's reading circuit.

This leads us to the present moment and the difficult, more specific questions that arise for children raised within a digital milieu, and ourselves. Will new readers develop the more time-demanding cognitive processes nurtured by print-based mediums as they absorb and acquire new cognitive capacities emphasized by digital media? For example, will the combination of reading on digital formats and daily immersion in a variety of digital experiences—from social media to virtual games—impede the formation of the slower cognitive processes such as critical thinking, personal reflection, imagination, and empathy that are all part of deep reading? Will the mix of continuously stimulating distractions of children's attention and immediate access to multiple sources of information give young readers less incentive either to build their own storehouses of knowledge or to think critically for themselves?

In other words, through no intention on anyone's part, will the increasing reliance of our youth on the servers of knowledge prove the greatest threat to the young brain's building of its own foundation of knowledge, as well as to a child's desire to think and imagine for him- or herself? Or will these new technologies provide the best, most complete bridge yet to ever more sophisticated forms of cognition and imagination that will enable our children to leap into new worlds of knowledge that we can't even conceive of in this moment of time? Will they develop a range of very different brain circuits? If so, what will be the implications of those different circuits for our society? Will the very diversity of such circuits benefit everyone? Can an individual reader consciously acquire various circuits, much like bilingual speakers who read different scripts?

Systematically examining—cognitively, linguistically, physiologically, and emotionally—the impact of various mediums on the acquisition and maintenance of the reading brain is the best preparation for ensuring the preservation of our most critical capacities both in the young and in ourselves. We need to understand the present expert brain's profoundly important cognitive contributions, as we add new cognitive and perceptual dimensions to its circuitry. No binary approach to either the formation or the preservation of the expert reading brain will be sufficient to meet the needs of the next generation or our own. The issues involved cannot be reduced simply to differences between print- and technology-based mediums. As the futurists Juan Enriquez and Steve Gullans wrote in *Evolving Ourselves: How Unnatural Selection and Nonrandom Mutation Are Changing Life on Earth*, we have choices to make in our evolution that will be more human-driven than nature-driven. These choices will be clear only if we stop to understand exactly what is involved with any important change. With you as my partner in dialogue, I seek to create within these letters a moment out of time to attend to the issues and choices we have before us, before the changes to the reader's brain are so ingrained that there is no going back.

Perhaps counterintuitively, I have chosen a rather odd, even anachronistic genre from the past, a book of letters, to address issues about a future that is changing moment by moment. I do so for reasons that spring from my experiences both as reader and as author. Letters invite a kind of cerebral pause in which we can think with each other and, if very fortunate, experience a special kind of encounter, what Marcel Proust called the "fertile miracle of communication" that occurs without ever moving from your chair. More specific to this genre, when I was young, Rainer Maria Rilke's *Letters to a Young Poet* influenced me greatly. As I grew older, however, it was not his lyrical language in those letters that touched me most, but the example of his consummate kindness toward an aspiring poet he had never met: Franz Xaver Kappus, a person whom he grew to care for only through letters. I have no doubt that both were changed through their exchanges. What better definition for a reader? What better model for an author? I hope the same for us.

Italo Calvino's *Six Memos for the Next Millennium* affected me similarly, even though his memos transcend any conventional notion of "letter" and were, to all our loss, unfinished. Both letters and memos are genres that bring Calvino's emphases on "lightness" to issues whose great weight might otherwise make their discussion too heavy for many to confront. Letters allow thoughts that, even when as urgent as some of the ones to be described, contain those ineffable aspects of lightness and connection that provide the basis of a true dialogue between author and reader—all accompanied by an impetus for new thoughts in you that will go in different directions from my own.

In a curious way, I have been involved in such a dialogue for some time. After I wrote *Proust and the Squid*, I received hundreds of letters from readers in every walk of life: famous literary figures concerned about their readers; neurosurgeons worried about their medical students in teaching hospitals in Boston; high school students forced to read a passage from my book on the Massachusetts state exam! It was heartwarming to me that the students were surprised to encounter my worry for their generation. Those letters showed me that what had begun as a book on the story and science of reading had become a cautionary tale about issues that have now become reality. The act of reflecting upon the major themes that my letter writers wrestled with prepared me for selecting the themes of each letter in this book and also for choosing this genre.

With this book I hope to go much further than I have in all my past work. That said, each letter will be informed by everything I have done before, particularly the research from my most recent articles and books, all found in the extensive notes at the end of the book that expand some of the issues encountered here. Letter Two is based on the largest body of that research, but it is also the most lighthearted of my letters to you, with its unapologetically whimsical overview of present knowledge about the reading brain. I hope to illuminate there both why the plasticity of the reading-brain circuit underlies the growing complexity of our thought and why and how this circuit is changing. In Letter Three I lead you into the essential processes that compose deep reading—from the reader's empathic and inferential abilities to critical analysis and insight itself. These first three letters provide a shared base from which to consider how the characteristics of various mediums, specifically print and screen reading, have begun to be reflected not only in the malleable networks of the brain's circuitry but also in *how and what* we now read.

The implications of our reading brain's plasticity are neither simple nor transient. The connections between how and what we read and what is written are critically important to today's society. In a milieu that continuously confronts us with a glut of information, the great temptation for many is to retreat to familiar silos of easily digested, less dense, less intellectually demanding information. The illusion of being informed by a daily deluge of eye-byte-sized information can trump the critical analysis of our complex realities. In Letter Four I confront these issues head-on and discuss how a democratic society depends on the undeterred use of these critical capacities and how quickly they can atrophy in each of us unnoticed.

In Letters Five to Eight, I morph into a "reading warrior" for the world's future children. Here I describe a range of concerns, from preserving the different roles that reading plays in their intellectual, social-emotional, and ethical formation, to worries about the vanishing aspects of childhood. Given their more particular worries, many parents and grandparents have asked me the equivalent of Kant's three questions: What do we know? What should we do? What can we hope? In Letters Six through Eight I provide a developmental proposal in which I describe my best thoughts about each of these questions, culminating in a rather unexpected plan for building a biliterate reading brain.

Toward that end, there will be no binary solutions proffered in any part of this book. One of the most important current outgrowths of my research involves working toward global literacy, where I publicly advocate and help in the design of digital tablets as one means of ameliorating nonliteracy, particularly for children with no schools or in inadequate ones. Do not think that I am against the digital revolution. Indeed, it is of utmost importance to be informed by the growing knowledge on the impact of different media if we are to prepare all our children, wherever they live, to read deeply and well, in whatever medium.

All of these letters will prepare you, my reader, to consider the many critical issues involved, beginning with yourself. In the last letter I ask you to think about who the true "good readers" are in our changing epoch and to reflect for yourself on the immeasurably important role they play in a democratic society—never more so than now. Within these pages the meanings of good reader have little to do with how well anyone decodes words; they have everything to do with being faithful to what Proust once described as the heart of the reading act, going beyond the wisdom of the author to discover one's own.

There are no shortcuts for becoming a good reader, but there are lives that propel and sustain it. Aristotle wrote that the good society has three lives: the life of knowledge and productivity; the life of entertainment and the Greeks' special relationship to leisure; and finally, the life of contemplation. So, too, the good reader. In the final letter I elaborate how this reader—like the good society—embodies each of Aristotle's three lives, even as the third life, the life of contemplation, is daily threatened in our culture. From the perspectives of neuroscience, literature, and human development I will argue that it is this form of reading that is our best chance at giving the next generation the foundation for the unique and autonomous life of the mind they will need in a world none of us can fully imagine. The expansive, encompassing processes that underlie insight and reflection in the present reading brain represent our best complement and antidote to the cognitive and emotional changes that are the sequelae of the multiple, life-enhancing achievements of a digital age.

Thus, in my last and most personal letter, you and I will face ourselves and ask whether we possess each of the three lives of the good reader, or whether, barely noted by us, we have lost the ability to enter our third life and, in so doing, have lost our reading home. Within that act of examination, I will suggest that the future of the human species can best sustain and pass on the highest forms of our collective intelligence, compassion, and wisdom by nurturing and protecting the contemplative dimension of the reading brain.

Kurt Vonnegut compared the role of the artist in society to that of the canary in the mines: both alert us to the presence of danger. The reading brain is the canary in our minds. We would be the worst of fools to ignore what it has to teach us.

You won't agree with me all the time, and that is as it should be. Like St. Thomas Aquinas, I look at disagreement as the place where "iron sharpens iron." That is my first goal for these letters: that they become a place where my best thoughts and yours will meet, sometimes clash, and in the process sharpen each other. My second goal is for you to have the evidence and information necessary to understand the choices you possess in building a future for your progeny. My third goal is simply what Proust hoped for each of his readers:

> It seemed to me that they would not be "my readers" but readers of their own selves, my book being merely a sort of magnifying glass. . . . I would furnish them with the means of reading what lay inside themselves.

Sincerely,
Your Author

Letter Two

UNDER THE BIG TOP
An Unusual View of the Reading Brain

The Brain—is wider than the Sky—
For—put them side by side—
The one the other will contain
With ease—and You—beside

The Brain is deeper than the sea—
For—hold them—Blue to Blue
The one the other will absorb—
As Sponges—Buckets—do—

The Brain is just the weight of God—
For—Heft them—Pound for Pound—
And they will differ—if they do—
As Syllable from Sound—
                              —Emily Dickinson

Dear Reader,

Emily Dickinson is my favorite nineteenth-century American poet. She was my favorite poet before I ever realized
how much she wrote about the brain, all from the most unlikely and circumscribed of observation posts, her second-
floor window on Main Street in Amherst, Massachusetts.

When she wrote "Tell all the truth, but tell it slant, Success in Circuit lies," she could never have known about the brain's many circuits. But like the great nineteenth-century neurologists, she had an intuitive understanding of the brain's "wider than the Sky" protean capacities: that is, the brain's quasi-miraculous ability to go outside its boundaries to develop new, never before imagined functions.

The neuroscientist David Eagleman recently wrote that the brain's cells are "connected to one another in a network of such staggering complexity that it bankrupts human language and necessitates new strains of mathematics. . . . there are as many connections in a single cubic centimeter of brain tissues as there are stars in the Milky Way galaxy." It is the capacity to make these mind-reeling numbers of connections that allows our brain to go beyond its original functions to form a completely new circuit for reading. A new circuit was necessary because reading is neither natural nor innate; rather, it is an unnatural cultural invention that has been scarcely six thousand years in existence. On any "evolutionary clock," reading's history occupies little more than the proverbial tick before midnight, yet this set of skills is so important in its capacity to change our brains that it is accelerating our species' development, for better and sometimes for worse.

Building a Reading Brain

It all begins with the principle of "plasticity within limits" in the brain's design. What amazes me most is not the brain's multiple sophisticated functions, but the fact that it is able to go beyond its original, biologically endowed functions—like vision and language—to develop totally unknown capacities such as reading and numeracy. To do so, it forms a new set of pathways by connecting and sometimes repurposing aspects of its older and more basic structures. Think about what an electrician does when asked to put new wiring into an old house to accommodate a contemporary, unplanned-for track lighting system. With no slight to the electrician, our brain goes about rewiring us in a much more ingenious way. Faced with something new to learn, the human brain not only rearranges its original parts (e.g., the structures and neurons responsible for essential functions such as vision and hearing), but it is also able to refit some of its existing neuronal groups in those same areas to accommodate the particular needs of the new function.

It is no coincidence, however, that the neuronal groups that are to be repurposed share similar functions with the new one. As the Parisian neuroscientist Stanislas Dehaene has noted, the brain recycles and even repurposes neuronal networks for skills that are cognitively or perceptually related to the new one. It is a wonderful example of our brain's plasticity within limits.

This ability to form newly recycled circuits permits us to learn all manner of genetically unplanned-for activities—from making the first wheel, to learning the alphabet, to surfing the net while listening to Coldplay and sending tweets. None of these activities is hardwired or has genes specifically dedicated to its development; they are cultural inventions that involve cortical takeovers. Nevertheless, there are significant and even difficult implications for the fact that reading is not hardwired the way language is.

In contrast to reading, oral language is one of our more basic human functions. As such, it possesses dedicated genes that unfold with minimal assistance to produce our capacities to speak and understand and think with words. In language, nature is nurtured by need in a fairly universal sequence around the world. This is why a young child, placed in any typical language environment, will learn to speak that language virtually without instruction. That is a wondrous thing.

Not so with newbie developments such as reading. To be sure, there are genes involved for basic capacities such as language and vision that become rearranged to form the reading circuit, but in and of themselves these genes do not produce the ability to read. We human beings have to learn to read. That means we must have an environment that helps us to develop and connect a complex assortment of basic and not-so-basic processes, so that every young brain can form its own brand-new reading circuit.

I want to underscore something essential here: with no genetic blueprint for reading, *there is no one ideal reading circuit.* There can be different ones. Unlike the development of language, the lack of a blueprint for reading circuitry means that its formation is subject to considerable variation, based on the reader's specific language requirements and learning environments. For example, a Chinese, character-based reading-brain circuit has both similarities to and discernible differences from an alphabet-reading brain. A large, fundamental mistake—with many unfortunate consequences for children, teachers, and parents around the world—is the assumption that reading is natural to human beings and that it will simply emerge "whole cloth" like language when the child is ready. This is not the case; most of us must be taught the basic principles of this unnatural cultural invention.

Happily, the brain comes well prepared to learn a great many unnatural things because of its basic design. The best-known design principle, *neuroplasticity,* underlies just about everything interesting about reading—from forming a new circuit by connecting older parts, to recycling existing neurons, to adding new and elaborated branches to the circuit over time. Most important for this discussion, however, plasticity also underlies why the reading-brain circuit is inherently malleable (read changeable) and influenced by key environmental factors: specifically, *what it reads* (both the particular writing system and the content), *how it reads* (the particular medium, such as print or screen, and its effects on the way we read), and *how it is formed* (methods of instruction). The crux of the matter is that the plasticity of our brain permits us to form both ever more sophisticated and expanded circuits and also ever less sophisticated circuits, depending on environmental factors.

The second principle invokes the contributions of the mid-twentieth-century psychologist Donald Hebb, who helped conceptualize how cells form working groups or cell assemblies, which help them to become specialists for particular functions. In reading, working groups of neuronal cells in each of the circuit's structural parts (such as vision and language) learn to perform some of the most highly specific functions. These specialist groups build the networks that allow us to see the smallest features of letters or hear the tiniest elements in the sounds of language, or *phonemes*, literally in milliseconds.

More specifically and equally important, cell specialization enables each working group of neurons to become automatic in its specific region and to become virtually automatic in its connections to the other groups or networks in the reading circuit. In other words, for reading to occur, there must be sonic-speed automaticity for neuronal networks at a local level (i.e., within structural regions like the visual cortex), which, in turn, allows for equally rapid connections across entire structural expanses of the brain (e.g., connecting visual regions to language regions). Thus, whenever we name even a single letter, we are activating entire networks of specific neuronal groups in the visual cortex, which correspond to entire networks of equally specific language-based cell groups, which correspond to networks of specific articulatory-motor cell groups—all with millisecond precision. Multiply this scenario a hundredfold when the task is to depict what you are doing when reading this very letter with complete (or even incomplete) attention and comprehension of the meanings involved.

In essence, the combination of these three principles forms the basis of what few of us would ever suspect: a reading circuit that incorporates input from two hemispheres, four lobes in each hemisphere (frontal, temporal, parietal, and occipital), and all five layers of the brain (from the uppermost telencephalon and adjacent diencephalon below it; to the middle layers of the mesencephalon; to the lower levels of the metencephalon and myelencephalon). Anyone who still believes the archaic canard that we use only a tiny portion of our brains hasn't yet become aware of what we do when we read.

Circuit du Soleil

If we as a society are to grapple with the full implications of the ongoing changes to our plastic reading brain, we need to get "under the hood" of the reading circuit. Or perhaps, with a little suspension of disbelief on your part, under the tent. To bring to life the multiple, simultaneously happening operations in the reading brain that occur every time we read a single word, I can think of no better visual metaphor than a three-ring circus. Not just any three-ring circus but one filled with actors and fantastic creatures only imaginable in a Cirque du Soleil tent where magic trumps credibility! With the help of the neuroscientist and gifted artist Catherine Stoodley, that is what I want you to experience.

FROM THE BIG TOP

Imagine yourself inside a round wooden perch at the very
top of a huge circus tent looking down on the scene below. From this vantage point, the formation of the reading
circuit resembles very much what goes on in the multiple
acts of a three-ring circus. But in our reading circus, there
will be five rings with ensembles of fantastically dressed
performers at the ready to act out the gamut of processes
necessary for us to read a single word. Fortunately for us
both, at my request, we are seeing only what happens in
the left hemisphere for now and, even more important, in
slow motion, so you can watch all that goes on without
being dizzied by the nearly automatic speeds that are involved.

Turn your attention first to the groups of performers
in the three large overlapping rings and then to the two
slightly smaller rings connected to the larger ones. Each
of the large rings depicts the expansive regions underlying
Vision, Language, and Cognition and represents one of the
original parts that are connected in the new reading circuit. The first of the two smaller rings represents the Motor
functions, whose performers are needed for the articulation of speech sounds and some other rather astonishing
activities that will unfold shortly. Unsurprisingly, this ring
is connected not only to Language but, more surprisingly,
also to Cognition. The other ring, which is linked to both
Language and Cognition, holds Affective functions and
connects the great range of our feelings to our thoughts
and words. Now turn your gaze to a lighted glass box on
the far left, where all manner of "very important persons"
seem to be executing their very important things. This box
is something akin to our brain's personal executive center, where various forms of attention, memory, hypothesis

generating, and decision making are carried out in an area right behind our foreheads called the prefrontal cortex.

Imagine that these major rings are superimposed over large structural regions that include the various layers of the brain (see Figure 1 for one of Stoodley's inimitable drawings of just the top, cortical layer of the reading brain). The ring of Vision takes up a great deal of the occipital lobe in the left hemisphere and some of the right hemisphere, at least for our alphabetic systems. Like the rings of Language and Cognition, the visual ring incorporates areas in the midbrain and the cerebellum to coordinate all its activities in almost automatic speeds. In contrast to the visual needs of the alphabetic reading system, the Chinese and Japanese Kanji writing systems use significantly more of the right hemisphere's visual regions to process all the visually demanding characters their readers need to remember and connect to concepts.

The ring of Language occupies expansive territory with regions in multiple layers in both hemispheres, particularly

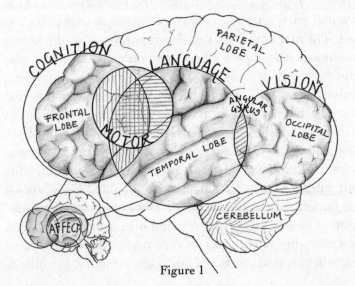

Figure 1

the parietal and temporal lobes adjacent to Vision, and also areas in the frontal lobe adjacent to the Motor areas. Similarly, the ring of Cognition and the deeper-situated ring of Affect (some of whose networks are formed down in the diencephalon, or second layer of the brain, right below the cerebral cortex) have considerable overlap with the Language areas.

The proximity and overlap of many parts of these rings are like a physical analogue to how closely aligned and interdependent their functions are. This view of the rings represents our first rudimentary glimpse of the reading circuit for the English writing system.

SPOTLIGHTS OF ATTENTION

Now let's look more closely at what happens inside the layers of the rings when we read a single word in English. As if on command, a huge image of a word that we can't yet make out properly is flashed across the largest top flap of the tent right below the level of our eyes. We have to move our attention quickly to follow the suddenly lit beams of several spotlights that were just turned on by the prefrontal control box. The brain's attentional systems are the equivalent of biological spotlights: unless the lights are turned on, nothing else can happen. But note that there are different kinds of spotlights. This is because the brain needs to be able to allocate different forms of attention to each of the many steps or processes involved in reading. What few people ever appreciate is how central attention is for every function that we perform and that multiple forms of attention go into action before our eyes even see the word.

The first spotlights, which do the work of the orienting attentional system, have three quickly accomplished jobs. First, they help us *disengage from* whatever we were originally attending to—which takes place in the parietal lobe of our cortex (i.e., in the telencephalon's uppermost layer). Second, they help *move our attention to* whatever is in front of us—in this case, to the specific word on the tent flap. This act of moving our visual attention takes place deep in our midbrain (i.e., in the mesencephalon, or third layer). Third, they help *focus* our new attention and, in so doing, alert the entire reading circuit to prepare for action. This last pre-reading focusing of attention takes place in a special area below the cortex that functions as one of the brain's major switchboards: the very important thalamus, which resides in the diencephalon, or second layer, of each hemisphere.

For the real action in the circuit to commence, however, we still need one more specific set of spotlights, organized by the prefrontal control box's executive attention system within both frontal lobes. This critical system manages whatever follows in a kind of cognitive workspace. Among other things, from the outset it holds our sensory information in working memory, so that we can integrate the different forms of information that are gathered there and not lose track of any of it. This is what allows you to do everything from solving math problems "in your head" to remembering digits in a phone number, letters in a word, and words in a sentence. There is an extremely tight relationship between the attentional system and the various kinds of memory.

RING OF VISION

After all this preliminary directing of our attention, something astonishing happens. The action we have been waiting for begins! Now moving rapidly out of the retinas are what appear to be two troupes of cyclists for each eye, made up of brightly clad actors on huge unicycles. These troupes are about to ride their cycles across the highest and longest of the wires that crisscross the entire length of the brain from the retinas in the eyes to the farthest point in the most posterior regions of the brain, the occipital lobes. The troupes in each eye begin together, but they quickly divide up at an X-shaped intersection called the *optic chiasm*, which appropriately enough looks like a junction of railroad tracks. At this juncture the four sets of troupes separate, with one set of the cyclists for each eye going on opposite paths through multiple layers of the brain to reach the visual areas in the back of both hemispheres. The way they are organized means that each eye sends one group of its cyclists to each hemisphere. It is a masterful design with great evolutionary advantages. Consider: even with only one eye, we have two hemispheres providing us with essential visual information.

The four groups of cyclists must make several stops along the way, but they seem undeterred as they carry their information at lightning speed. Within 50 milliseconds they all arrive with their messages at one very specific area in the occipital lobes called the *visual striate cortex*, which gets its name from the stripes created by its six layers of alternating white and gray matter.

After their arrival in the fourth layer of this cortical region, the cyclists fan out (see Figure 2). Suddenly the whole of the ring of Vision in the occipital lobes is awhirl with activity. The information from the high-wire cyclists is rapidly transferred to clusters of tiny orblike creatures that look vaguely . . . well, like little eyes with arms and feet. One group of these industrious orbs identifies the cyclists' message as a set of "letters" and immediately passes that information over to neighboring orblike creatures in deeper regions of the cortex that signal that they are real, permissible letters. Another group quickly examines the features that make up the letters (e.g., lines, circles, and diagonals) and identifies them as the well-known English letters *t+r+a+c+k+s*.

It would seem almost immediately upon the second group's recognition of the letters in the word that multiple acts are performed by other teams of specialized neurons that burst into action. Some orbs perform only on single letters, while others respond to the letter patterns found in words, such as the *ack* and *tr* in *tracks*; others identify the most commonly used, meaningful parts of words called *morphemes* (e.g., prefixes and suffixes such as the plural *s* in our word). It becomes clear that each working group in this ring has its own territorial domain and works quickly and skillfully on just those highly specific bits of visual information. We can't help but notice that some groups of orbs appear placidly uninterested or at least underemployed, with little activation upon seeing our word. Some of them identify only the most frequently seen whole words such as *stop* and *the*—words often called *sight words* that need no further analysis by other visual neurons. Others are obviously dedicated to other visual features.

What is anything but obvious is how the bicycle artists locate with such rapid precision the exact groups of neuronal orbs that are poised to identify their particular bits of visual information. Perhaps unsurprising by now, behind this mystery lies another set of remarkable design principles—in this case, *retinotopic organization* and *representation*. In retinotopic organization, highly differentiated neurons in the retina trigger particular corresponding neurons in the visual areas. Not unlike having their own GPS system, the cyclists' rapid-fire ability to locate the right neurons facilitates their extremely precise and swift transfer of information. In the case of letters,

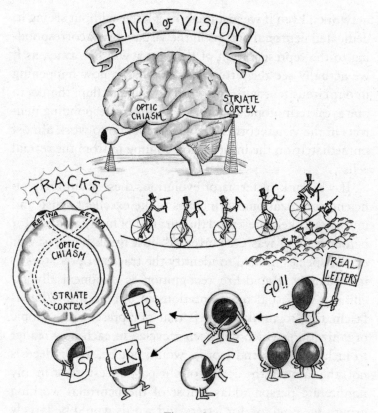

Figure 2

the retinal troupes have to learn to make these connections through multiple exposures in a long developmental process.

This learning is facilitated by the brain's ability to make representations (think re-presentations) of patterns like letters. The visual cortex of an expert reader is chock-full of representations of letters, as well as common letter patterns and word parts (such as the morphemes that make up the roots, prefixes, and suffixes of our words) and even many well-known words. It is hard to imagine at first, but these representations possess a physical reality in our neuronal networks. Even if we just imagine a letter without seeing it, dedicated neuronal groups in the visual cortex corresponding to the representation of that letter will fire away, as if we actually see the letter. This is what is now happening in our circus tent with the word on the tent flap: thanks to our eyes' retinotopic organization, the corresponding neurons in the visual cortex are already set up to work almost immediately on the information coming in from the retinal cells.

If we think in terms of evolution, these stunningly efficient organizational principles make excellent sense and more than likely preserved the survival of many an ancestor before reading was ever invented. Just think how fast our earlier species needed to identify the tracks of predators—immediately. Rapid-fire recognition is exponentially facilitated by visual representations in our brains. What is fascinating to think about is that our present retinotopic organization, which has been recycled in each new reader to include letters and words, would not be, and indeed is not, the same in the cortex of our past ancestors or in any nonliterate person today. Most of the neuronal working groups we now use for letters and words would be largely dedicated in nonliterate individuals to visually similar but functionally different tasks, such as the identification of objects or faces. This is a prime example of how, when the brain learns to read, it repurposes some networks originally used to identify the small features within objects and faces to recognize the similarly small features in letters and words.

RING OF LANGUAGE

But now we need to get back to the circus. Right on cue, we notice several startling events as new troupes of neurons from the ring of Language begin to spring into motion, the operative word being *spring*. All number of flying, gyrating actors are leaping around the area of the ring of Language that borders Vision, where the occipital and temporal lobes meet. To be sure, it will take many neuronal groups to ensure, first, that the visual information (i.e., the letters) is quickly connected with the correct sound- or phoneme-based information in our word, and, second, that this information is connected to all the word's possible meanings and associations.

English has around forty-four different phonemes (depending on the dialect used), represented here by forty-four exceedingly small actors jumping around impatiently in the dynamically expanding Language ring. Like ponies in a racing stall, the diminutive performers are poised for the moment when some of them will be linked to their visual partners in *t+r+a+c+k+s*. We notice that some clusters of actors look a lot like Siamese twins or triplets. They are the ones responsible for common sounds that blend together, such as the *tr* in our unfolding word. It also appears that the most frequently used sounds are given an advantage in their place in the ring, as if in anticipation of being chosen first in any matchup process.

There is a reason for this. Just off to the left, in our peripheral vision, we see how the control box seems to be highlighting the most likely probabilities of which letters or letter groups are to be chosen. Clearly nothing about the expert reading brain is left to chance, but rather is based on probabilities and prediction that, in turn, are based on context and prior knowledge. After this initial guidance from the prefrontal areas, pandemonium breaks out as the corresponding phoneme actors match the sounds corresponding to the visual troupes' input. The word *tracks* is a go, and the fireworks commence!

Delight is palpable throughout the tent, with whole new groups of performers in the rings of Language and Cognition joining the act. Somersaulting acrobats plunge in front of the word, each shouting all manner of interesting possible meanings: "animal tracks, sport tracks, railway tracks?" The acrobats are mesmerizing in their flexibility, moving from one possible, frequent meaning to less frequently used words and on to a new litany of other possibilities: "tracks of tears, audio tracks, school tracks, one-track mind, eye tracking, track lighting?"

As if these semantic-based meanings were not enough, grinning mimes straddling the Language and Motor rings ask, "What about the verb *tracks*?," proposing yet further possibilities. There is something akin to a collective intake of breath from a previously unknown sector in the adjacent Motor ring. There, a lively set of exotically costumed mimes appears at the ready, poised to articulate the word or, far more mysteriously, physically act it out. Without obviously moving the close-by neurons controlling the muscles of the lips, larynx, and tongue, they are preparing to simulate moving the muscles of legs and hands, depending on whether the meaning of the word is an action verb or more abstract: "tracks an animal, tracks a crime, tracks data trends, tracks a hurricane."

Behind the whirling acrobats and mimes can be seen in the wings hundreds of other acrobat and mime groupings, all in the same "semantic neighborhood." A few of them stand very close to the ring, prepared to leap inside with related words and concepts on millisecond notice. There are groups of words primed to perform simply because they sound like *tracks* in alliterative ways (e.g., treats, trams, trains, tricks), or possess possible rhymes (e.g., packs, sacks, lacks, and even wax).

RING OF COGNITION AND RING OF AFFECT

And, as if to raise our attention away from the solo per-
former acts in the Language ring, brilliantly clothed flying
trapeze artists leap overhead to one another, lifting our consciousness to completely different, remembered thoughts in
the huge unexplored spaces now beckoning us to enter overlapping regions in the ring of Cognition. As the trapeze figures drift in and out, we hear them whisper questions to us
about contexts for the word *tracks* that we had not thought
about at first glance. There is a scene from childhood in
which a tiny train is panting up railway tracks that lead up
and down imposing hills, muttering, "I think I can, I think I
can." Another little train on very similar tracks is bright blue
and called Thomas the Tank Engine. In yet another scene,
huge muscled men are splitting logs to build train tracks in
what looks like nineteenth-century America (see Figure 3).

Feelings of childhood begin to rise in us along with these
images, and with them the ring of Affect begins to pulsate
with different feelings associated with the active thoughts
and words in the other rings. But not only feelings from
childhood are aroused; a group of actors is becoming increasingly visible on the other side of the ring of Cognition.
Gradually we see that they are an ensemble of winter-clothed
people staring in horror at the figure of a beautiful Russian
woman with long black hair and a red bag: it is Anna Karenina, about to throw herself across a set of *tracks*! But just as
familiar feelings of dread, empathy, and sadness rise from
the ring of Affect, the scene dims and our attention moves.

A very unusual, almost phantasmal apparition now appears, perched above an area called the *angular gyrus*. The
location of this region at the juncture of the occipital, temporal, and parietal lobes is pivotal and reflects its ability to

Figure 3

integrate functions from the ring of Vision in the occipital lobe and the rings of Language and Cognition in the temporal and parietal lobes (see Figure 1). The large, tuxedo-clad figure doesn't speak and appears to be something between a ringmaster and a train switchmaster, integrating information and selecting the tracks of the words for us to follow.

Whether commands come from this figure or from the prefrontal control box or both is unclear, but the lights

on the ring of Cognition now dim, and the wraithlike figure of Anna fades from our view and consciousness. There
was not enough information to stay with Anna's image,
even if we are left with an ever-so-slight frisson of vestigial
sadness and regret. We realize in this instant that there is
always something that remains within us from all previous encounters with this seemingly ordinary word, *tracks*,
and indeed with many words. Just as the cognitive scientist
David Swinney underscored years ago, our words contain
and momentarily activate whole repositories of associated
meanings, memories, and feelings, even when the exact
meaning in a given context is specified.

Within this millisecond of recollection, we begin to appreciate the multilayered beauty in the brain's design for storing
and retrieving words: each word can elicit an entire history of
myriad connections, associations, and long-stored emotions.
Indeed, you have just witnessed how the reading brain activates
in half a second something akin to the daily efforts of poets and
writers to find the perfect word, the mot juste, that will connect, as E. M. Forster once described it, "prose with passion."

Let us finish our tour of the reading brain by looking
down one final time at all that we have seen in our imag-
inary *circuit de la lecture*. This time, however, I have ar-
ranged for you to see the action not in slow motion but in
real time—in little more than 400 milliseconds—and across
both hemispheres. With almost impossible rapidity, we can
now see that at the start of things the right-hemisphere visual areas quickly cross over to the left, where all manner
of activation occurs and is integrated across all layers of the
rings. Finally, at the end of the action we see that much of
the right hemisphere is now alight with multiple areas contributing to the meanings of *tracks*, fewer for its sounds. We
can't perceive more than that. Our eye simply cannot follow
the movements in the acts quickly enough to comprehend

exactly what is happening where and when. In fact, the scene now looks like a seamless performance of such intimately connected networks that the image we are left with appears to be one huge set of pulsating, connected lights. Indeed, "there are as many connections" in the reading brain's circuitry "as there are stars in the Milky Way galaxy."

This final image of the reading brain's connectedness conveys that at least as many things are happening in zigzagging, feed-forward, and feed-backward interactivity as are occurring linearly. Indeed, such an impression would be the closest approximation of the many unknowns that remain about the timing and sequence of all that goes on between and among the rings of Vision, Language, Cognition, Motor, and Affect when we read. We are left at the top of our circus tent humbled by the enormity of what makes up this reading act—which most humans take completely for granted.

I hope you never will. Rather, I hope that you now understand that whenever you read a single word, you activate thousands upon thousands of neuronal working groups, all the ones you just encountered and many more. And if we activate myriads of neurons with just one word, imagine what you bring to bear when you read a sentence of multiple words, an essay by Nicholas Kristof, a poem by Adrienne Rich, short stories by Andrea Barrett, a book on language by Ray Jackendoff, a work of literary criticism by Michael Dirda. After all the years of research I have pursued to understand what we do when we retrieve a single word, I still remain in awe of what happens when we read a line of words that elicits our deepest thoughts.

As will be discussed in the next letter, the deep-reading brain very literally, physiologically goes "everywhere" to understand. But that could change.

Sincerely yours,
Your Author

DEEP READING
Is It Endangered?

I think that reading, in its original essence, [is] that fertile miracle of communication effected in solitude. . . . We feel quite truly that our wisdom begins where that of the author leaves off. . . . But by a singular and moreover providential law . . . (a law which perhaps signifies that we are unable to receive the truth from anyone else but must create it ourselves), that which is the endpoint of their wisdom appears to us as but the beginning of our own.

—Marcel Proust, *On Reading*

Dear Reader,

You have just "tracked" the pathways of a single word. We saw in the last letter that the reading of a single word invokes the activation of myriads of neurons, which involves the transmission of signals across multiple regions and all five layers of the brain. Now imagine that rather than reading the single word *tracks* in isolation, I ask you to decode and comprehend this word in the much more demanding context of a sentence, such as this next one.

His love left no tracks, save for the kind that never go away—for her and any who would follow.

What's in a Sentence?

If I ever write a novel, I will fill it with sentences like this, which just asked a great deal more of you than met your eye. If my Tufts colleagues Gina Kuperberg and Phillip Holcomb were to use their various brain-imaging techniques, from fMRI to event-related potential techniques (ERPs), to watch your brain as you finished reading this sentence, you would be able to observe remarkable expanses of processes that were required to comprehend the varied possible, even surprising meanings conveyed by it. For example, after you encountered the word *tracks* in this context, you would see in the ERPs what is called an N400 response in several language-based regions. The brain-wave activity at around 400 milliseconds in these areas gives an electrophysiological signal of your brain's surprise. These regions have registered something anomalous and unanticipated—in this case, a meaning that was not initially predicted about the word *tracks*, particularly after all the different meanings of *tracks* were just primed or activated in you during the last letter. Sentences in which our initial predictions of the meaning of a word are not confirmed require a cerebrally pregnant pause, especially if, as in this haunting sentence, we are to understand the poignant inferences the last words quietly direct us toward. In sentences such as these, the whole is far greater than the sum of the parts, and the reading-brain circuit reflects this in which processes are activated, how long, and where.

Your processing of this sentence, or indeed any sentence, is not a simple additive exercise in which all the perceptual and linguistic activities described earlier in the rings now take place for twenty words in a row. As Andy Clark persuasively writes, when we read words in sentences and longer text, we enter new cognitive territory, where prediction meets perception, and indeed, more often than not, precedes and prepares perception. It is still a matter of amazement to me that what we know before we read any sentence prepares us to recognize even the visual shapes of the individual words faster and to understand their meanings more rapidly and more precisely in any new context. We who are expert readers process and connect our lower-level perceptual information (i.e., the first rings of the reading circuit) at near-breakneck speeds. Only such speeds can enable us to allocate attention to the higher-level deep-reading processes, which in turn constantly feed their conclusions back and forth with the lower-level processes, thus better preparing them for the next words they encounter.

The cognitive beauty of these interactive exchanges is that they accelerate everything from perception to comprehension. They accelerate perception by narrowing the possibilities of what we will read next to a set of words that correspond to what Gina Kuperberg calls "proactive" predictions. It's what every smartphone is now doing as you type your words, if occasionally with wild (and sometimes embarrassing) misses. These predictions in turn stem from various sources, including our working memory of what we have just read and our longer-term memory of stored background knowledge. Together, these interactions among perception, language, and deep-reading processes accelerate our understanding because they allow us to read a sentence of twenty words as a sum of predicted thoughts far more quickly than the sum of information provided by twenty individually read words.

The quality of how we read any sentence or text depends, however, on the choices we make with the time we allocate to the processes of deep reading, regardless of medium. Everything we will consider in this book from this

point forward—from the digital culture, the reading habits
of our children and their children to the role of contemplation in ourselves and in society—rests on understanding
the critically important but never guaranteed allocation of
time to the processes that form the deep-reading circuit.
This is true of both the development of the circuit in childhood and its maintenance over the course of our lives. It
takes years for deep-reading processes to be formed, and as
a society we need to be sure that we are vigilant about their
development in our young from a very early age. It takes
daily vigilance by us, the expert readers of our society, to
choose to expend the extra milliseconds needed to maintain deep reading over time.

A LITTLE WINDOW INTO YOURSELF

Let's see how well you are doing just that.

Consider these two passages from the famous geneticist
Francis S. Collins, the director of the Human Genome Project, about reading the best-known text ever written, the Bible.

---

*Find a Bible right now and read Genesis 1:1 through
Genesis 2:7. There is no substitute for looking at the
actual text if one is trying to understand its meaning.*

*Despite twenty-five centuries of debate, it is fair
to say that no human knows what the meaning of
Genesis 1 and 2 was precisely intended to be. We
should continue to explore that! But the idea that
scientific revelations would represent an enemy
in that pursuit is ill conceived. If God created the
universe, and the laws that govern it, and if He endowed human beings with intellectual abilities to*

DEEP READING   39

*discern its workings, would He want us to disregard
those abilities?*

Chances are good that you read Collins's first passage
about the two accounts of Creation in Genesis 1 and 2 by
reading quickly and with little effort. The second passage,
however, may have given you more than one pause. Nevertheless, the chances are better than good that you read it in one
of two very different ways: either with considerable effort to
attend to and reflect upon what Collins meant about science
and religious belief when reading Genesis; or on the fly of
your attention. The way you read these two passages provides
a milliseconds-long window not only into your own current
reading, but also into dilemmas we all face in this new millennium, as we move from a literacy- and word-based culture
into a far faster-paced digital- and screen-based one.

In one of his poems, William Stafford wrote, "a quality
of attention has been given you." It was a poet's description
of the cognitive strata below the surface of words that invite
us to discover thoughts that are to be glimpsed nowhere else.
It is the nature of attention—which you just used to explore
or to skim Francis Collins's words—that underlies large,
unanswered questions that society is beginning to confront.
Will the quality of our attention change as we read on mediums that advantage immediacy, dart-quick task switching,
and continuous monitoring of distraction, as opposed to the
more deliberative focusing of our attention?

What concerns me as a scientist is whether expert readers
like us, after multiple hours (and years) of daily screen reading, are subtly changing the allocation of our attention to key
processes when reading longer, more demanding texts. Will
our quality of attention in reading—the basis of the quality
of our thought—change inexorably as our culture transitions

away from a print-based culture toward a digital one? What are the cognitive threats to and the promises of such a transition? To understand what we may be losing and what we may gain as we acquire and use the skills necessary for daily life in the twenty-first century, I want to dive straight to the heart of the matter, through an examination of the manifold varieties of the deep-reading processes that make up the expert reading-brain circuit, so that their diverse capacities speak for themselves. The deep-reading processes described here are not meant as an exclusive list, nor do they appear in the brain in any single sequence or configuration. Some are more evocative in their function, some more analytical, some more generative. Depending on the type of reading, multiple complex processes activate in dynamic tandem in the reading-brain circuit, with input from one another and, as alluded to, with the word-level input from before. Regardless of their sequence, as the old Chinese manservant told his young charges in John Steinbeck's *East of Eden*, "at the end there is light."

Deep Reading's Evocative Processes

> When we reflect that "sentence" means, literally, "a way of thinking" . . . we realize that . . . a sentence is both the opportunity and the limit of thought—what we have to think with, and what we have to think in. It is, moreover, a *feelable* thought. . . . It is a pattern of felt sense.
>
> —Wendell Berry

IMAGERY

Wendell Berry's conceptualization of the sentence as a "*feelable* thought" is a good segue into one of deep reading's most tangible, sensorially evocative processes: our capacity to form images when we read. How do we do this? As the artist-writer Peter Mendelsund emphasized, what we "see" when we are reading helps us to co-create images with the author or sometimes, as in some fiction, through the author's surrogate. It is similarly the case for the voice of the narrator we hear in both fiction and nonfiction. As one novelist describes this handoff, "Open a book and a voice speaks. A world, more or less alien or welcoming, emerges to enrich a reader's store of hypotheses about how life is to be understood." Thus, when you read Mark Twain's description of Huckleberry Finn, Alice Walker's depiction of Celie, or F. Scott Fitzgerald's use of Nick Carraway's voice to describe Jay Gatsby, you could almost pick each of those characters out in a crowd. Together you and the author construct images out of a set of carefully chosen, sensory details conveyed only by words.

Take one of the most compelling "short short stories" ever written. It emerged as the result of a wager made to Ernest Hemingway by his unruly group of writing friends. They bet him that he couldn't write a story in six words. It is hardly surprising that Hemingway took and won the bet. The surprise is that he felt that this story was one of his finest pieces of writing. He was right. With a bare minimum of words, he evoked one of the most powerful visual images, and also some of the same deep-reading processes we might utilize when reading his longer works. Here is his story in six words:

For sale: baby shoes, never worn.

Few examples of just six words have ever delivered such a visceral punch. We feel with intuitive certainty why the shoes were never worn. Before this realization, you will have seen in your mind's eye the image of a lonely pair of baby shoes, probably with perfect, diminutive laces and no hint of a tiny foot's imprint. Such an image will have given sad entry into your reservoir of background knowledge that helped you infer an entire scenario beneath the surface information of the sale. Simultaneously, the interactions within your own background knowledge, imagery, and inferential processes helped you to move from your own perspective into the perspective of others, with all the mix of emotions that might add.

Thus, in six terse words Hemingway presented an image capable of giving the reader a range of personal emotions: a wrenching sense of the feelings such a loss would bring; a barely suppressed relief at not having had the experience, with the sting of guilt that follows such a sense of relief; and, perhaps, a prayerlike hope never to know the feeling more intimately. Few writers could ever make us enter this mix of desperate feelings through such an economy of words. But it is not Hemingway's journalist-influenced economy in writing that is my focus here, but rather the capacity of imagery to help us both enter the multiple layers of meaning that can underlie text and also understand the thoughts and feelings of others.

Empathy: "Passing Over" into the Perspective of Others

> Only connect.
> —E. M. Forster

The act of taking on the perspective and feelings of others is one of the most profound, insufficiently heralded contributions of the deep-reading processes. Proust's description of "that fertile miracle of communication effected in solitude" depicts an intimate emotional dimension within the reading experience: the capacity to communicate and to feel with another without moving an inch out of our private worlds. This capacity imparted by reading—to leave and yet not leave one's sphere—is what gave the reclusive Emily Dickinson what she called her personal "frigate" to other lives and lands outside her perch above Main Street in Amherst, Massachusetts.

The narrative theologian John S. Dunne described this process of encounter and perspective taking in reading as the act of "passing over," in which we enter into the feelings, imaginings, and thoughts of others through a particular kind of empathy: "Passing over is never total but is always partial and incomplete. And there is an equal and opposite process of *coming back* to oneself." It is a beautifully apt description for how we move from our inherently circumscribed views of the world to enter another's and return enlarged. In *Love's Mind*, his numinous book on contemplation, Dunne expanded Proust's insight: "That 'fruitful miracle of a communication effected in solitude' may be already a kind of *learning to love*." Dunne saw the paradox that Proust described within reading—in which communication occurs despite the solitary nature of the reading act—as an unexpected preparation for our efforts to come to know other human beings, understand what they feel, and begin to change our sense of who or what is "other." For theologians such as John Dunne and writers such as Gish Jen, whose lifework illumines this principle in fiction and nonfiction alike, the act of reading is a special place in which human beings are freed from themselves to pass over to others and, in so doing, learn what it means to be another person with aspirations, doubts, and emotions that they might otherwise never have known.

A powerful example of the transforming effects of

"passing over" was told to me by a Berkeley-trained drama teacher who works with adolescents in the heart of the Midwest. A student came to him, a beautiful thirteen-year-old girl, who said she wanted to be part of his theater group performing William Shakespeare's plays. It would have been an ordinary request, save for the reality that the young girl had advanced cystic fibrosis and had been told she had only a brief time to live. That amazing teacher gave the young girl a role he hoped would give her the feelings of romantic love and passion that she might never experience in life. She became, he said, the perfect Juliet. Almost overnight, she memorized the lines of *Romeo and Juliet* as if she had played the role a hundred times before.

It was what happened next that stunned everyone around her. She went on to become one Shakespearean heroine after another, each role performed with more emotional depth and strength than the one before. Years have now passed since she played Juliet. Against all expectations and medical prognoses, she has entered college, where she is pursuing a dual degree in medicine and theater, in which she will continue to "pass over" into one role after another.

That young woman's exceptional example is not so much about whether the mind and heart can overcome the limitations of the body; rather, it is about the powerful nature of what entering the lives of others can mean for our own lives. Drama makes more visible what each of us does when we pass over in our deepest, most immersive forms of reading. We welcome the Other as a guest within ourselves, and sometimes we become Other. For a moment in time we leave ourselves; and when we return, sometimes expanded and strengthened, we are changed both intellectually and emotionally. And sometimes, as this remarkable young woman's example shows us, we experience what life has not allowed us. It is an incalculable gift.

And there is a gift within a gift. Perspective taking not only connects our sense of empathy with what we have just read but also expands our internalized knowledge of the world. These are the learned capacities that help us become more human over time, whether as a child when reading *Frog and Toad* and learning what Toad does when Frog is sick or as an adult when reading Toni Morrison's *Beloved*, Colson Whitehead's *Underground Railroad*, or James Baldwin's *I Am Not Your Negro,* and experiencing the soul-stealing depravity of slavery and the desperation of those condemned to it or to its legacy.

Through this consciousness-changing dimension of the act of reading, we learn to feel what it means to be despairing and hopeless or ecstatic and consumed with unspoken feelings. I no longer remember how many times I have read what each of Jane Austen's heroines felt—Emma, Fanny Price, Elizabeth Bennet in *Pride and Prejudice* or in her newest incarnation in Curtis Sittenfeld's *Eligible: A Modern Retelling of Pride and Prejudice.* What I know is that each of those characters experienced emotions that helped me understand the range of the often contradictory feelings each of us possesses; doing so leaves us feeling less alone with our particular complex mix of emotions, whatever our life's circumstances. As expressed in the play *Shadowlands*, about the life of C. S. Lewis, "We read to know that we are not alone."

Indeed if we are very lucky, we may come to experience a special form of love for those who inhabit our books and even, at times, for the authors who write them. One of the most concrete renderings of this latter concept can be found in the most unlikely of historical persons, Niccolò Machiavelli. In order that he might better enter the consciousness and "converse" with the authors he was reading, he would dress formally in the style of dress appropriate to the authors in their various epochs. In a letter to the diplomat Francesco Vettori in 1513, he wrote:

> I am not ashamed to speak with them, and to ask them the reasons for their actions; and they in their kindness answer me; four hours may pass and I do not feel boredom, I forget every trouble, I do not dread poverty, I am not frightened by death; I give myself entirely to them.

In this passage Machiavelli exemplifies not only the perspective-taking dimension of deep reading, but also the capacity to be transported from whatever our present realities are to an internal place where we can experience a sharing of the inevitable burdens that typify most human existence whatever our age: fear, anxiety, loneliness, sickness, love's uncertainties, loss and rejection, sometimes death itself. I do not doubt that some of this was what the young Susan Sontag felt when she would look at her bookcase and feel she was "looking at my fifty friends. A book was like stepping through a mirror. I could go somewhere else." And surely it is what these authors give witness to in the communicative dimension of reading and what it means at every age to leave oneself to enter the welcome solace of the company of others, whether fictional characters, historical figures, or the authors themselves.

That this freely given immersion in the reading life could be threatened in our culture has begun to emerge as a concern for growing numbers in our society, including an NPR team that spent a whole interview with me on their personal concern about this loss. There are many things that would be lost if we slowly lose the *cognitive patience* to immerse ourselves in the worlds created by books and the lives and feelings of the "friends" who inhabit them. And although it is a wonderful thing that movies and film can do some of this, too, there is a difference in the quality of immersion that is made possible by entering the articulated thoughts of others. What will happen to young readers who never meet and begin to understand the thoughts and feelings of someone totally different? What will happen to older readers who begin to lose touch with that feeling of empathy for people outside their ken or kin? It is a formula for unwitting ignorance, fear and misunderstanding, that can lead to the belligerent forms of intolerance that are the opposite of America's original goals for its citizens of many cultures.

Such thoughts and their correlative hope are frequent themes in the work of the novelist Marilynne Robinson, whom former president Barack Obama described as a "specialist in empathy." In one of the most remarkable of exchanges requested by him during his presidency, Obama visited Robinson on a trip to Iowa. During their wide-ranging discussion, Robinson lamented what she saw as a political drift among many people in the United States toward seeing those different from themselves as the "sinister other." She characterized this as "dangerous a development as there could be in terms of whether we continue to be a democracy." Whether writing about humanism's decline or fear's capacity to diminish the very values its proponents purport to defend, she conceptualizes the power of books to help us understand the perspective of others as an antidote to the fears and prejudices many people harbor, often unknowingly. Within this context, Obama told Robinson that the most important things he had learned about being a citizen had come from novels: "It has to do with empathy. It has to do with being comfortable with the notion that the world is complicated and full of grays but there's still truth there to be found, and that you have to strive for that and *work* for that. And the notion that it's possible to connect with someone else even though they're very different from you."

The desperately real lessons about empathy that Obama and Robinson discussed may begin with the experiencing of other lives, but they are deepened by the work that follows perspective taking—when something we read forces us to examine our own prior judgments and the lives of others. Lucia Berlin's story "A Manual for Cleaning Women" is a case in point for me. When I began the story, I saw the protagonist cleaning woman as being oblivious to the everyday tragedies that skirted just below the surface in the places where she worked. Until, that is, I read the last sentence, which ended the story with her utterance "I finally weep." Everything I had first assumed about the cleaning woman narrator in this story collapsed with the final line. My false and circumscribing inferences flew out one of those windows that open when we see the prejudices we bring to whatever we read. No doubt that was the humbling realization that Berlin intended her readers to discover about themselves.

James Carroll's book *Christ Actually: The Son of God for the Secular Age* describes a similar confrontation with perspective taking in the realm of nonfiction. There he related his experiences as a young, very devout Catholic boy reading *Anne Frank: The Diary of a Young Girl*. He described the life-changing epiphany he had felt upon entering the life of that young Jewish woman with all her undiminished young girl's hopes and enthusiasm for life, all of which she sustained despite the violent hatred of Jews that ultimately destroyed her and her family.

Entering the perspective of this completely foreign girl provided an unexpected rite of passage for the young James Carroll. From his memorable descriptions of his conflicts with his military general father during the Vietnam crisis in *An American Requiem: God, My Father, and the War*

That Came Between Us to his descriptions of the relationship between Judaism and Christianity in *Constantine's Sword: The Church and the Jews: A History*, each of his books revolves around the need to understand, at the deepest level, the perspective of the *other*, whether in Vietnam or in a German concentration camp.

In *Christ Actually*, he used the life and thought of the early-twentieth-century German theologian Dietrich Bonhoeffer to underscore the life-and-death consequences of human failure to take on the perspective of other. Bonhoeffer preached and wrote unflinchingly, first from a pulpit and then from a prison cell, about the tragic inability of most people at the time both to understand the perspective of the historical Jesus as a Jew and to see the persecution of Jews in Germany from their perspective. At the heart of his last work, he asked: How would the historical Christ actually respond to Nazi Germany? Only he who shouts for the Jews, he asserted, can "sing their Gregorian chants." That conclusion led him to act against his own religious beliefs about murder by contributing to two unsuccessful attempts on Hitler's life and ultimately to being killed in a concentration camp on direct orders from the Führer's representative.

I write this letter during a time when millions of refugees—most of whom are Muslim—are fleeing horrific conditions and trying to enter Europe, the United States, or anywhere else they can to regain their previous lives. I write this letter on the day a young Jewish boy from my own city of Boston has been killed in Israel during his gap year before college because he was perceived by a young Palestinian boy as the "enemy other." Developing the deepest forms of reading cannot prevent all such tragedies, but understanding the perspective of other human beings can give ever fresh, varied reasons to find alternative, compassionate ways to deal with the others in our world, whether they are innocent Muslim children crossing treacherous open seas or an innocent Jewish boy from Boston's Maimonides School, all killed miles and miles away from their homes.

The unsettling reality, however, is that unbeknownst to many of us, including until recently myself, there has begun an unanticipated decline of empathy among our young people. The MIT scholar Sherry Turkle described a study by Sara Konrath and her research group at Stanford University that showed a 40 percent decline in empathy in our young people over the last two decades, with the most precipitous decline in the last ten years. Turkle attributes the loss of empathy largely to their inability to navigate the online world without losing track of their real-time, face-to-face relationships. In her view our technologies place us at a remove, which changes not only who we are as individuals but also who we are with one another.

Reading at the deepest levels may provide one part of the antidote to the noted trend away from empathy. But make no mistake: empathy is not solely about being compassionate toward others; its importance goes further. For it is also about a more in-depth understanding of the Other, an essential skill in a world of increasing connectedness among divergent cultures. Research in the cognitive neurosciences indicates that what I call perspective taking here represents a complex mix of cognitive, social, and emotional processes that leaves ample tracks in our reading-brain circuitry. Brain-imaging research by the German neuroscientist Tania Singer expands former conceptualizations of empathy to show that it involves a whole feeling-thinking network that connects vision, language, and cognition with extensive subcortical networks. Singer emphasizes that this larger network comprises, among other areas, the highly connected neuronal networks for theory of mind, including the insula and the cingulate cortex, which function to connect large expanses of the human brain. Often undeveloped in many individuals on the autism spectrum and lost in a pathological condition called *alexithymia*, theory of mind refers to an essential human capacity that allows us to perceive, analyze, and interpret the thoughts and feelings of others in our social interactions with them. Singer and her colleagues describe how the very large neurons in these areas are uniquely suited for the extremely rapid communication necessary in empathy between these areas and other cortical and subcortical regions, including, of all places, the motor cortex.

Though it may seem something of a figurative leap to think that the motor cortex is activated when you read, it is closer to a literal, cortical hop. Reconstruct the fleeting image evoked in the last letter with the image of Anna Karenina leaping upon the tracks. For those of you who read that passage in Tolstoy's novel, *you leaped, too.* In all likelihood the same neurons you deploy when you move your legs and trunk were also activated when you read that Anna jumped before the train. A great many parts of your brain were activated, both in empathizing with her visceral despair and in some mirror neurons acting this desperation out motorically.

Although mirror neurons may have become more popular than they are fully understood, they play a fascinating role in reading. In what is surely one of the more intriguingly titled articles in this research, "Your Brain on Jane Austen," the scholar of eighteenth-century literature Natalie Phillips teamed with Stanford neuroscientists to study what happens when we read fiction in different ways: that is, with and without "close attention." (Think back to the two Collins quotes.) Phillips and her colleagues found that when we read a piece of fiction "closely," we activate regions of the brain that are aligned to what the characters are both feeling and doing. She and her colleagues were frankly surprised that just by asking their literature graduate students either to read closely or to read for entertainment, different regions of the brain became activated, including multiple areas involved in motion and touch.

In related work, neuroscientists from Emory University and from York University have shown how networks in the areas responsible for touch, called the somatosensory cortex, are activated when we read metaphors about texture, and also how motor neurons are activated when we read about movement. Thus, when we read about Emma Bovary's silken skirt, our areas of touch are activated, and when we read about Emma stumbling from her carriage to run in pursuit of Léon, her young, fickle lover, areas responsible for motion in our motor cortex activate, and, more than likely, those in many affective areas do, too.

These studies are the beginning of increasing work on the place of empathy and perspective taking in the neuroscience of literature. The cognitive scientist Keith Oatley, who studies the psychology of fiction, has demonstrated a strong relationship between reading fiction and the involvement of the cognitive processes known to underlie both empathy and theory of mind. Oatley and his York University colleague Raymond Mar suggest that the process of taking on another's consciousness in reading fiction and the nature of fiction's content—where the great emotions and conflicts of life are regularly played out—not only contribute to our empathy, but represent what the social scientist Frank Hakemulder called our "moral laboratory." In this sense, when we read fiction, the brain actively simulates the consciousness of another person, including those whom we would never otherwise even imagine knowing. It allows us to try on, for a few moments, what it truly means to be another person, with all the similar and sometimes vastly different emotions and struggles that govern others' lives. The reading circuitry is elaborated by such simulations; so also our daily lives, and so also the lives of those who would lead others.

The novelist Jane Smiley worries that it is just this dimension in fiction that is most threatened by our culture: "My guess is that mere technology will not kill the novel. . . . But novels can be sidelined. . . . When that happens, our society will be brutalized and coarsened by people . . . who have no way of understanding us or each other." It is a chilling reminder of how important the life of reading is for human beings if we are to form an ever more realized democratic society for everyone.

Empathy involves, therefore, both knowledge and feeling. It involves leaving past assumptions behind and deepening our intellectual understanding of another person, another religion, another culture and epoch. In this moment in our collective history, the capacity for compassionate knowledge of others may be our best antidote to the "culture of indifference" that spiritual leaders such as the Dalai Lama, Bishop Desmond Tutu, and Pope Francis describe. It may also be our best bridge to others with whom we need to work together, so as to create a safer world for all its inhabitants. In the very special cognitive space within the reading-brain circuit, pride and prejudice can gradually dissolve through the compassionate understanding of another's mind.

This emerging work on empathy in the reading brain illustrates physiologically, cognitively, politically, and culturally how important it is that feeling and thought be connected in the reading circuit in every person. The quality of our thought depends on the background knowledge and feelings we each bring to bear.

Background Knowledge

> Who is each one of us, if not a combinatoria of experiences, information, books we have read. . . . Each life is an encyclopedia, a library.
>
> —Italo Calvino

Many first-grade readers might be able to decode Hemingway's six-word story, but they would not have the background knowledge to infer its underlying meaning or to feel any of the emotions you and I experience when reading it. Over the life span, everything we read adds to a reservoir of knowledge that is the basis of our ability to comprehend and to predict whatever we read.

By reservoir, I do not refer only to facts, though they are surely part of it. Some of our finest writers have written eloquently about the conceptual building blocks that the reading of books has given their lives. In his beautiful work *A History of Reading*, Alberto Manguel exemplifies this essential component of deep reading when he writes that reading is cumulative. As a teenager, Manguel worked in the bookstore Pygmalion in Buenos Aires. There he encountered Pygmalion's most illustrious customer, Argentina's famed writer Jorge Luis Borges, who frequented the store to find not only new works but new readers. Borges had begun losing his sight in his fifties and would hire one person after another from the bookstore to read to him. The story of how Manguel became Borges's reader is one of the more touching accounts of two esteemed writers, one world famous, one yet to have written his first public words. What Manguel learned in Borges's personal library permeates every book he would go on to write, from *A Reader on Reading* to *The Library at Night*: that is, the profound impact of books upon the lives and knowledge stores of those who read them.

Both the work and the personal lives of Manguel and Borges give us portraits of the inestimable importance of the unique background knowledge that comes to us from what we read. I am concerned about both *what* we read and *how* we read. Does the content of what we are reading in our present milieu provide us sufficient background knowledge both for the particular demands of life in the twenty-first century and for the formation of the deep-reading brain circuit? We seem to be moving as a society from a group of expert readers with uniquely personal, internal platforms of background knowledge to a group of expert readers who are increasingly dependent on similar, external servers of knowledge. I want to understand the consequences and costs of losing these uniquely formed internal sources of knowledge without losing sight of the extraordinary gifts of the abundant information now at our fingertips.

Albert Einstein said that our theories of the world determine what we see. So also in reading. We must have our own wheelhouse of facts to see and evaluate new information, whatever the medium. If the brilliant futurist Ray Kurzweil is correct, it may be possible to have all those external sources of information and knowledge implanted within the human brain, but at present this is technologically, physiologically, and ethically not an option. For now, our internal background knowledge is as essential to the rest of deep reading as salt was to King Lear's pork and, perhaps, as little appreciated till it begins to disappear. The relationship between what we read and what we know will be fundamentally altered by too early and too great a reliance on external knowledge. We must be able to use our own knowledge base to grasp new information and interpret it with inference and critical analysis. The outline of the alternative is already clear: we will become increasingly susceptible human beings who are more and more easily led by sometimes dubious, sometimes even false information that we mistake for knowledge or, worse, do not care one way or another.

An answer to such scenarios is before our eyes: in the reciprocal relationship between background knowledge and deep reading. When you read carefully, you are more able to discern what is true and to add to it what you know. Ralph Waldo Emerson described this aspect of reading in his extraordinary speech "The American Scholar": "When the mind is braced by labor and invention, the page of whatever book we read becomes luminous with manifold allusion. Every sentence is doubly significant." In reading research, the cognitive psychologist Keith Stanovich suggested something similar some time ago about the development of word knowledge. In childhood, he declared, the word-rich get richer and the word-poor get poorer, a phenomenon he called the "Matthew Effect" after a passage in the New Testament. There is also a Matthew-Emerson Effect for background knowledge: those who have read widely and well will have many resources to apply to what they read; those who do not will have less to bring, which, in turn, gives them less basis for inference, deduction, and analogical thought and makes them ripe for falling prey to unadjudicated information, whether fake news or complete fabrications. *Our young will not know what they do not know.*

Others, too. Without sufficient background knowledge, the rest of the deep-reading processes will be deployed less often, leading to a situation in which many people will never move outside the boundaries of what they already know. For knowledge to evolve, we need to continuously add to our background knowledge. Paradoxically, most factual information today comes from external sources that can be unadjudicated and without proof of any form.

How we analyze and use this information and whether we cease to deploy the time-consuming, critical processes to evaluate new information will significantly impact our future. Absent the checks and balances provided by both our prior knowledge content and our analytical processes, we run the risk of digesting information without questioning whether the quality or prioritization of the information available to us is accurate and free from external motivations and prejudices.

We need to ensure that human beings do not fall into the trap that Edward Tenner described when he said, "It would be a shame if brilliant technology were to end up threatening the kind of intellect that produced it." At a recent conference, the director of the University of Alberta's library system, Gerald Beasley, spoke about the effects of the digital transition on the fate of books: "The present situation is unresolvable. Until it is, we must be the 'guardians of the book's attributes.'" The same is true for guarding the reader's attributes, which begin and end with what the reader knows.

In one of the most famous statements about scientific breakthroughs, Louis Pasteur wrote, "Chance comes only to the prepared mind." That elegant statement could just as easily describe the role of background knowledge in the deep-reading brain. It is an appropriate segue from how we bring a prepared mind to what we read to how we utilize our more analogical skills to analyze the information we construct and how we use that sifted thought as the stuff of whole new thoughts and insight.

To prepare you for these next processes, I'll end this section with a little lightness, through another "short short story" from the science fiction author Eileen Gunn. Her six words, ostensibly about space travel, may require a few extra STEM cells . . .

Computer, did we bring batteries?

Computer . . .

Deep Reading's Analytical Processes

Without concepts there can be no thought, and without analogies there can be no concepts . . . analogy is the fuel and fire of thinking.

—Douglas Hofstadter and Emmanuel Sander

It is hardly coincidental that what we think of as the methods of science characterize many of the most sophisticated cognitive processes we deploy during deep reading. Getting to the truth of things—whether in science, in life, or in text—requires observation, hypotheses, and predictions based on inference and deduction, testing and evaluation, interpretation and conclusion, and when possible, new proof of these conclusions through their replication. During the first milliseconds of reading we gather together what we perceive, integrating our observations. Analogical reasoning, as the cognitive scientist Douglas Hofstadter has written, provides the great bridge between what we see and what we know (background knowledge) and drives us to form new concepts and hypotheses. These hypotheses help guide the application of inferential capacities, such as deduction and induction, and in due course lead to the evaluation and critical analysis of what we think our observations and inferences mean. From these, we draw interpretations of all that went before and, if most fortunate, come to conclusions that lead to bursts of insight. There are both poetry and science at the heart of reading.

Just which of the scientific methods are deployed largely depends on the reader's expertise and the content being read. If we are reading a scientific article on mirror neurons in the motor system by the neuroscientist Leonardo Fogassi from Parma, for instance, we will need to evaluate whether the concepts, hypotheses, and findings presented build on past evidence; whether verifiable methods of evaluation are used that can be replicated; and whether the conclusions and interpretations match the data provided. In the process, we use a veritable armamentarium of analogical, inferential, and analytical processes, and we learn a great deal from Professor Fogassi that adds to our future background knowledge.

ANALOGY AND INFERENCE

On the other hand, if we are reading a poem by Wallace Stevens or an essay by the contemporary philosopher Mark Greif in *Against Everything*, we might well use different forms of inference, as well as a more nuanced range of emotions, than when we are reading . . . well, about motor neurons. Reading, at least all deep reading, requires the use of analogical reasoning and inference if we are to uncover the multiple layers of meaning in what we read. In beginning one of Greif's philosophical excavations into what we do in our ordinary lives and why, I confronted the obvious and less obvious reasons why I exercise. I wouldn't want to spend a minute in the gym of Greif's observations, where the grunts, moans, and pyrotechnics of wannabe Amazons and Prometheans could make anyone be against anything they are doing. But I don't look at exercise that way, which is Greif's beautifully, subversively rendered point: to use an examination of our most basic activities and motives to get us to think about what we do with our "one wild and precious life."

Greif's rants, only seemingly against everything, are a

powerful example of how analogical thought and inferential reasoning help us comprehend what lies beneath the surface of the increasingly complex world he examines. The more we know, the more we can draw analogies, and the more we can use those analogies to infer, deduce, analyze, and evaluate our past assumptions—all of which increases and refines our growing internal platform of knowledge. The converse is equally true, with harsh implications for our present and future society: the less we know, the fewer possibilities we have for drawing analogies, for increasing our inferential and analytical powers, and for expanding and applying our general knowledge.

Sherlock Holmes provides a masterful example of how careful observation, background knowledge, and analogical reasoning lead to deductions that continue to astonish us. At the core of our perennial fascination with Sir Arthur Conan Doyle's master sleuth is the mesmerizing way in which Holmes derives brilliant inferences from the most prosaic sources: two short (rather than long) brown dog hairs on the right pants leg and a tiny set of not-yet-healed scratches on the left hand; a hint of moisture still apparent under the lapel; a one-way ticket stub at 4:00 p.m. from Cambridge to London. Voilà! The disheveled professor with the corner of the damp ticket stub still visible in his breast pocket is now the prime suspect. He lied three times: first, about whether he had been near the rainy Cambridge train station; second, about his whereabouts at 4:00, the hour of the murder; and third, about whether he had recently seen the unfortunate victim and her equally unlucky brown short-haired Jack Russell terrier (a species that can bark incessantly and loudly, the probable cause of its sad decease).

Holmes's methods, which are the basis of one mystery series after another on both sides of the Atlantic, echo our own inferential abilities even though his are fictional and ours have something extra. Unlike Holmes (particularly his brilliant asocial portrayal by Benedict Cumberbatch), and more like the perspicacious Miss Marple, we often combine inferential capacities with empathy and perspective taking to ferret out the mysteries in what we read.

Our brains favor Miss Marple. Widely distributed networks in our left and right prefrontal cortex analyze the text's information and then make predictions that go into a kind of internal peer review system to evaluate the worth of each hypothesis. Indeed, some research indicates that the left prefrontal region connects observations and inferences and then makes one self-generated hypothesis after another. Meanwhile, the right prefrontal cortex evaluates the worth of each prediction and then sends this judgment back to the left prefrontal area for the final imprimatur. It is akin to watching the scientific method in action but with the addition of the networks for empathy and theory of mind thrown in the solutions. In the long run, the deep-reading brain's more mixed-methods use of analogical, inferential, and empathy-related processes is ultimately to be preferred over Sherlock's. It's a deduction!

The consistent strengthening of the connections among our analogical, inferential, empathic, and background knowledge processes generalizes well beyond reading. When we learn to connect these processes over and over in our reading, it becomes easier to apply them to our own lives, teasing apart our motives and intentions and understanding with ever greater perspicacity and, perhaps, wisdom, why others think and feel the way they do. Not only is it the basis for the compassionate side of empathy, but it also contributes to strategic thinking.

Just as Obama noted, however, these strengthened processes do not come without work and practice, nor do they remain static if unused. From start to finish, the basic neurological principle—"Use it or lose it"—is true for each deep-reading process. More important still, this principle holds for the whole plastic reading-brain circuit. Only if we continuously work to develop and use our complex analogical and inferential skills will the neural networks underlying them sustain our capacity to be thoughtful, critical analysts of knowledge, rather than passive consumers of information.

CRITICAL ANALYSIS

Such a statement leads inevitably to the key integrative role of critical analysis in the deep-reading circuit. Whether from the perspective of science, education, literature, or poetry, more has been written about critical thinking than about any other of the deep-reading processes because of its pivotal place in intellectual formation. Nevertheless, critical analysis remains as difficult to define as to foster. From the standpoint of the reading brain, critical thought represents the full sum of the scientific-method processes. It synthesizes the text's content with our background knowledge, analogies, deductions, inductions, and inferences and then uses this synthesis to evaluate the author's underlying assumptions, interpretations, and conclusions. The careful formation of critical reasoning is the best way to inoculate the next generation against manipulative and superficial information, whether in text or on screen.

That said, in a culture that rewards immediacy, ease, and efficiency, the demanding time and effort involved in developing all the aspects of critical thought make it an increasingly embattled entity. Most of us think we are exercising critical thinking, but if we are honest with ourselves, we realize that we are doing so less than we imagine. We believe we will allocate time to it "later," that invisible wastebasket of lost intentions.

In the literary scholar Mark Edmundson's laudable book *Why Read?*, he asks, "What exactly is critical thinking?" He explains that it includes the power to examine and potentially debunk personal beliefs and convictions. Then he asks, "What good is this power of critical thought if you do not yourself believe something and are not open to having these beliefs modified? What's called critical thought generally takes place from no set position at all."

Edmundson articulates here two connected, insufficiently discussed threats to critical thinking. The first threat comes when any powerful framework for understanding our world (such as a political or religious view) becomes so impenetrable to change and so rigidly adhered to that it obfuscates any divergent type of thought, even when the latter is evidence-based or morally based.

The second threat that Edmundson observes is the total absence of any developed personal belief system in many of our young people, who either do not know enough about past systems of thought (e.g., contributions by Sigmund Freud, Charles Darwin, or Noam Chomsky) or who are too impatient to examine and learn from them. As a result, their ability to learn the kind of critical thinking necessary for deeper understanding can become stunted. Intellectual rudderlessness and adherence to a way of thought that allows no questions are threats to critical thinking in us all.

Critical thought never just happens. Years ago, my family and I were taken by the Israeli philosopher Moshe Halbertal to see a school in Mea She'arim, the deeply Orthodox area of Jerusalem to which we would otherwise never have been invited to go. Halbertal's work on ethics and morality permeates his profoundly thoughtful, sometimes controversial approaches to some of the most difficult political and spiritual issues facing our contemporary world, including the Israel Defense Forces' Code of Ethics, which he helped write. I looked through the windows of that school to see young boys rocking, praying, singing, and arguing with one another back and forth about the possible interpretations of single lines of text in the Torah. No one interpretation was assumed; rather, an entire history of commentaries was to be brought to bear on those often bare lines of text. These young readers were expected to use their understanding of past knowledge—in this case, centuries of exchanged thought—as a point of departure for their own.

Something akin to this form of intellectual analysis occurs within the deepest forms of reading, wherein different possible interpretations of text move back and forth, integrating background knowledge with empathy and inference with critical analysis. Critical analysis in its deepest forms, therefore, represents the best possible integration of past hard-sought thoughts *and* feelings, which is the single best preparation for a whole new understanding. In the wonderful ways in which words can reveal concepts beyond themselves, such a mode of critical thinking is at once a catholic, Talmudic bridge to new thought.

Deep Reading's Generative Processes

An insight is a fleeting glimpse of the brain's huge store of unknown knowledge. The cortex is sharing one of its secrets.
—Jonah Lehrer

We arrive at last at the end of the reading act. Insight is the culmination of the multiple modes of exploration we have brought to bear on what we have read thus far: the information harvested from the text; the connections to our best thoughts and feelings; the critical conclusions gained; and then the uncharted leap into a cognitive space where we may upon occasion glimpse whole new thoughts. As the philosopher Michael Patrick Lynch tells it, "Realization comes in a flash. . . . Insight . . . is the opening of a door, a 'disclosing' as Heidegger said. One acts by opening the door, and then one is acted upon by seeing what lies beyond. Understanding is a form of disclosure."

The evanescent nature of what happens when we experience deep insights makes their impressions no less lasting. I invite you to pause for a moment and reflect upon some of your most important insights as you read this last section. To prime your memories, I give you three examples from different stages of my reading life—two from fiction, one from science. My first example comes once again from Rilke—not his *Letters* but the most unlikely collection of tales in his *Stories of God*. I read those tender tales as a twenty-year-old and have never forgotten the delicate insights I gained from them. In one story, a group of children has been carefully taking turns carrying and guarding something that they solemnly believe is God. That is, until the smallest child loses "It." Day turns into night, the other children leave, and only the smallest child remains, searching desperately and futilely. She asks all passersby if they can help her try to find God, but no one can. At last, when all hope seems gone, a stranger suddenly appears. Leaning down to her, he says he doesn't know where to find God, but he has just discovered a little thimble on the ground.

I still remember the frisson of pure joy I felt when the child held "God" safely back in her hand. I saw with what tenderness Rilke had woven together his thoughts about childhood belief and how the tiny thimble gave new form to the infinitely various ways in which we human beings try to "hold on to" God. I also realized how many insights reach us, just as Shakespeare had Polonius tell us, through indirection, which leads us more slowly, and perhaps more unerringly, to the sweetest of aha's.

More recently, Marilynne Robinson's *Gilead* has become for me a book of insights that change each year along with me. In this quiet story set in a place and time where nothing seems to happen, the Reverend John Ames labors to write a set of letters and remembrances to his very young son that will preserve and pass on the wisdom of his older generation long after the gentle curate is gone. Few works of fiction so deftly portray some of the most difficult, unresolvable questions about belief in God, an afterlife, forgiveness, virtue, and the miracle that we exist at all. This loving man's efforts to give the shoulders of his life's thoughts to his small son point us toward one of the most loving functions of insight within deep reading: to leave our best thoughts for those who will follow.

One of my favorite quotes about insight and creative thoughts comes from my third example, an article in *Psychological Bulletin* by the neuroscientists Arne Dietrich and Riam Kanso that reviews what is known from brain-imaging studies about insight and creative thought. In what is as close as scientists get to expressing exasperation in a peer-reviewed publication, they conclude, "It might be stated that creativity is everywhere." Despite poring over multiple EEG, ERP, and other neuroimaging studies, they could find no neat map of what occurs when we have our most creative bursts of thinking. Rather, it appears that we activate multiple regions of the brain, particularly the prefrontal cortex and the anterior cingulate gyrus (mentioned earlier in many other deep-reading processes that involve empathy, analogy, analysis, and their connections). Such findings are not so much exasperating as perhaps the perfect description of the great host of processes that converge when we individual readers generate a single, new thought and render it within what Wendell Berry lovingly described as the opportunity and limits of the sentence.

There is fact and there is mystery when we humans enter the last milliseconds of reading the sentences before us. Whether we use the beautiful metaphor of a "holding-ground for the contemplation of experience" by the literature scholar Philip Davis, the more psychological term of a "neuronal workspace" by the neuroscientist Stanislas Dehaene, or what the novelist Gish Jen called the reader's multichambered "interiority," there is a final moment in the reading act when an arms-wide expanse in the reader's mind opens up and all our cognitive and affective processes become the stuff of pure attention and reflection. Cognitively and physiologically, this pause is not a quiet or static time. It is an intensely active moment that can lead us even deeper into insights from the text or beyond them, as we sift past perceptions, feelings, and thoughts in pursuit of what the psychologist William James thought of and Philip Davis described as "that invisible generative place. . . . the invisible presence of mind behind, within, and between its words." Although it feels semi-sacrilegious to amend their thoughts, I would add "the invisible presence of the mind that reads behind, within, and between the words."

Novelists, philosophers, and neuroscientists present us with different slants on these last generative moments. However we conceptualize Emerson's "quarry" of language and thought, each reader of this book knows what is to be found there: the inestimable thoughts that from time to time irradiate our consciousness with brief, luminous glimpses of what lies outside the boundaries of all we thought before. In such moments, deep reading provides our finest vehicle to travel outside the circumferences of our lives.

Figure 4

The formation of the reading-brain circuit is a unique epigenetic achievement in the intellectual history of our species. Within this circuit, deep reading significantly changes what we perceive, what we feel, and what we know and in so doing alters, informs, and elaborates the circuit itself. Catherine Stoodley's final drawing of the reading brain illustrates how beautifully elaborated the deep-reading circuit becomes. As the next letter describes, however, the implications of the reading brain's plasticity make its future iterations in a digital milieu a matter of great consequence—and no small uncertainty.

Sincerely yours,
Your Author

Letter Four

"WHAT WILL BECOME OF THE READERS WE HAVE BEEN?"

> In common things that round us lie
> Some random truths he can impart, .
> —The harvest of a quiet eye.
> —William Wordsworth

> As the devotion of a life, the way of words, of knowing and
> loving words, is a way to the essence of things, and to the
> essence of knowing too. . . . What is required for a loving
> that is knowing, for a knowing that is loving, is the *quiet eye.*
> —John S. Dunne

Dear Reader,

At the end of "A Poet's Epitaph," William Wordsworth described the legacy that the poet brings to the world as the "harvest of a quiet eye." The artist Sylvia Judson used "quiet eye" to describe what she wants the viewer to bring to art. The theologian John Dunne used "quiet eye" to describe what humans need to suffuse love with knowledge. Contemporary golfers use the term to describe a method of improving their concentration; I wonder if professional golfers realize the poetry behind their swings.

I employ "quiet eye" to crystallize both my worries and

my hopes for the reader of the twenty-first century—whose eye increasingly will not stay still; whose mind darts like a nectar-driven hummingbird from one stimulus to another; whose "quality of attention" is slipping imperceptibly with consequences none could have predicted. In the last two letters you saw how the concentration of attention allows us to hold a word, a sentence, a passage in such a manner that we can move through multiple processes to all the layers of meaning, form, and feeling that enhance our lives. But what if our capacity to perceive is actually decreasing because we are confronted with too much information, as the philosopher Josef Pieper once wrote? What if we have become virtually addicted to the heightened sensory stimulation that composes much of our daily lives and cannot stop ourselves from pursuing it incessantly, as Judith Shulevitz suggests in *The Sabbath World: Glimpses of a Different Order of Time* and as technology experts in "persuasion design" principles know very well? The present letter to you confronts two central questions with implications far beyond any answer available now: Are we as a society beginning to lose the quality of attention necessary to give time to the essential human faculties that make up and sustain deep reading? If the answer is yes, what can we do?

Addressing these questions begins by understanding the fundamental tension between our evolutionary wiring and contemporary culture. Frank Schirrmacher, the late editor of the influential *Frankfurter Allgemeine Zeitung*'s *Feuilleton*, placed the origins of the conflict within our species' need to be instantly aware of every new stimulus, what some call our *novelty bias*. Hypervigilance toward the environment has important survival value. Undoubtedly this reflex saved many of our prehistoric ancestors from threats signaled by the barely visible tracks of deadly tigers or the soft susurrus of venomous snakes in the underbrush.

As Schirrmacher described it, the problem is that contemporary environments bombard us constantly with new sensory stimuli, as we splice our attention across multiple digital devices most of our days and, as often as not, nights shortened by our attention to them. A recent study by Time Inc. of the media habits of people in their twenties indicated that they switched media sources twenty-seven times an hour. On average they now check their cell phone between 150 and 190 times a day. As a society, we are continuously distracted by our environment, and our very wiring as hominids aids and abets this. We do not see or hear with the same quality of attention, because we see and hear too much, become habituated, and then seek still more.

Hyperattention is one of the inevitable by-products of this confluence. The literary critic Katherine Hayles characterized hyperattention as a phenomenon caused by (and then adding to the need for) rapid task switching, high levels of stimulation, and a low-level threshold of boredom. As early as 1998, Linda Stone, then part of the Virtual Worlds Group at Microsoft, coined the term *continuous partial attention* to capture the way children attend to their digital devices and then to their environments. Since that time, these devices have multiplied in number and ubiquity, including for the very young. A quick glance around you on your next plane trip will provide sufficient data for this observation. The iPad is the new pacifier.

There are unseen costs for every age. By a calculus we largely neglect, the more constant the digital stimulation, the more prevalent the boredom and ennui expressed by even very young children when we take the devices away. Further, the more the devices are used, the more dependent the entire family becomes on longer periods of access to digital sources of entertainment, information, and distraction by all its members. Hyperattention, continuous partial attention, and what the psychiatrist Edward Hallowell calls environmentally induced attentional "deficits" pertain to us all. From the minute we awaken to the alarm on one digital device, through attention-switching checks in fifteen-minute-or-less intervals on multiple other devices throughout the day, to the last minutes before we sleep when we perform our final, "virtuous" sweep of email to prepare for the next day, we inhabit a world of distraction.

There is neither the time nor the impetus for the nurturing of a quiet eye, much less the memory of its harvests. Behind our screens, at work and at home, we have sutured the temporal segments of our days so as to switch our attention from one task or one source of stimulation to another. We cannot but be changed.

And we are—in ways you have begun to sense. Over the last ten years we have changed in *how much* we read, *how* we read, *what* we read, and *why* we read with a "digital chain" that connects the links among them all and extracts a tax we have only begun to tally.

The Digital Chain Hypothesis

HOW MUCH WE READ

How much we read is a story in progress. Not long ago, the Global Information Industry Center at the University of California, San Diego, conducted a major study to determine the amounts of information we use daily and found that the average person consumes multiple gigabytes across varied devices each day. Basically, that is the equivalent of between 50,000 and 100,000 words a day. Asked to comment in an interview, a co-author of the study, Roger Bohn, is quoted as saying, "I think one thing is clear: our attention is being chopped into shorter intervals and that is probably not good for thinking deeper thoughts."

We can all appreciate the monumental efforts that went into this type of study, and its authors deserve great approbation. That said, "probably not good for thinking deeper thoughts" is maddeningly understated. Neither deep reading nor deep thinking can be enhanced by the aptly named "chopblock" of time we are all experiencing, or by 34 gigabytes of anything per day. To be sure, there are many thoughtful readers (and writers) such as James Wood who are reassured by the fact that we are all reading more, not less. After all, a little over a decade ago a report from the National Endowment for the Arts (NEA) raised a legitimate concern that many people were reading less than they had only a short time before, possibly due to the influence of digital reading. A few years later another report, initiated by the esteemed poet and then NEA director Dana Gioia, indicated that the trend had been reversed and that as a society we were reading more than ever, possibly spurred on by the same digitally based factor.

It is easy to be confused by our reading habits over the last years of transition from a literacy-based to a more digitally influenced culture. Whether based on the reports by the NEA or more recently updated ones, the reality at this point is that we have become so inundated with information that the average person in the United States now reads daily the same number of words as is found in many a novel. Unfortunately, this form of reading is rarely continuous, sustained, or concentrated; rather, the average 34 gigabytes consumed by most of us represent one spasmodic burst of activity after another. Little wonder that American novelists such as Jane Smiley worry that the novel, which requires and rewards a special form of sustained reading, will be "sidelined" by the ever-increasing barrage of words we feel compelled to consume daily. Writing in the 1930s, the German philosopher Walter Benjamin summarized the more universal dimension within this preoccupation with new information in a way that is equally, if not more, true today. We doggedly "pursue a present," he wrote, that consists of "information that does not survive the moment in which it is new."

From a reading researcher's perspective or, surprisingly enough, from the perspective of a former president of the United States, the kind of "information" that Benjamin described does not represent knowledge. The journalist and writer David Ulin quoted a speech by Barack Obama to students at Hampton University in which he worried that for many of our young, information has become "a distraction, a diversion, a form of entertainment, rather than a tool of empowerment, rather than the means of emancipation."

Obama's concern is one that is shared by increasing numbers of academic colleagues. The literature professor Mark Edmundson has written at length about the effects of his students conceptualizing information as a form of being entertained:

> Swimming in entertainment, my students have been sealed off from the chance to call everything they've valued into question, to look at new ways of life. . . . For them, education is knowing and lordly spectatorship, never the Socratic dialogue about how one ought to live one's life.

It is a message that speaks to the loss of both critical thinking and Proust's "communication in the midst of solitude," where the quiet eye of the reader must be still enough to hear the author, much less converse. Such an internal dialogue requires of the reader both time and desire. Edmundson worries that there is a diminishing desire among our young people to expend such effort, particularly if the alternative is to be passively entertained using the barely skimmed surface of their cognitive capabilities.

Edmundson's worries align with what knowledge about the reading circuit cautions: that if information is continuously perceived as a form of entertainment at the surface level, it remains on the surface, potentially impeding real thinking, rather than deepening it. Recall the imaging study in which the brains of Natalie Phillips's literature students showed less activation for casual reading than for deep or close reading. Reading light becomes one more entertaining distraction, however cleverly "masquerading as being in the know," as Ulin notes. Whether from Ulin's perspective as a journalist, a president's perspective as a protector of the nation's youth, or Edmundson's perspective as a teacher of young adults, the last thing a society needs is what Socrates feared: young people thinking they know the truth before they ever begin the arduous practice of searching for it.

As each of you, my readers, realizes, these worries can no longer be directed only toward our young. The sheer amount of information that we all consume involves one set of inherently game-changing issues after another. What do we do with the cognitive overload from multiple gigabytes of information from multiple devices? First, we simplify. Second, we process the information as rapidly as possible; more precisely, we read more in briefer bursts. Third, we triage. We stealthily begin the insidious trade-off between our need to know with our need to save and gain time. Sometimes we outsource our intelligence to the information outlets that offer the fastest, simplest, most digestible distillations of information we no longer want to think about ourselves.

And just as in many a translation from one language to another, things go missing—from the slackened use of our own individual analytic powers to a culture in which complex ideas are no longer the dominant currency. When we retreat from the intrinsic complexity of human life for whatever reason, often as not we turn to what conforms to the narrowing confines of what we already know, never shaking or testing that base, never looking outside the boundaries of our past thought with all its earlier assumptions and sometimes dormant but ready-to-pounce prejudices. We should know that we are making a Faustian bargain in our overamped lives and that unless we attend to what we are choosing—however unconsciously—we may lose, very literally, more than we think. We have already begun to change how we read—with all of its many implications for how we think, the next link in the chain.

HOW WE READ

> To be a moral human being is to pay, be obliged to pay, certain kinds of attention. . . . The nature of moral judgments depends on our capacity for paying attention—a capacity that, inevitably, has its limits, but whose limits can be stretched.
>
> —Susan Sontag

The story of how our reading is changing is hardly finished. Ziming Liu, Naomi Baron, Andrew Piper, David Ulin, and Anne Mangen's group in Europe, scholars from disparate disciplines and countries, are addressing how the type of reading on screens that we are now accustomed to is changing the very nature of our reading. Few would dispute what the information science and reading researcher Liu finds: that "skimming" is the new normal in our digital reading. Liu and various eye-movement researchers have described how digital reading often as not involves an *F* or zigzag style in which we rapidly "word-spot" through a text (often on the left-hand side of the screen) to grasp the context, dart to the conclusions at the end, and, only if warranted, return to the body of the text to cherry-pick supporting details.

Some of the most important questions about the effects of such a skimming style focus upon whether there are differences in using and maintaining higher-level reading comprehension processes. Naomi Baron's excellent meta-analysis of research on this indicates a mixed picture with regard to overall comprehension. Some of the most compelling studies concern changes in readers' grasp of the sequence of a plotline's details and possibly of the logical structure of an argument. The Norwegian scholar Anne Mangen is investigating cognitive and affective differences in print and screen reading in a program of research conducted with her colleagues Adriaan van der Weel, Jean-Luc Velay, Gerard Olivier, and Pascal Robinet. Mangen and her group asked student subjects to read and answer questions about a short story whose plot was deemed likely to have universal student appeal (a lust-filled French love story!). Half of the students read *Jenny, Mon Amour* on a Kindle, the other half in a paperback book.

The results indicated that the students who read the book were superior to their screen-reading peers in their ability to reconstruct the plot in chronological order. In other words, the sequencing of the sometimes easily overlooked details in a fictional story appeared to be lost by the students who were reading on a digital screen. Think what would happen in the stories of O. Henry if you skimmed the details—such as the wife cutting and selling her hair to buy the watch fob for her husband while he was selling his beloved watch to give her a comb for her beautiful hair. What is hypothesized by Mangen and a growing group of researchers is that their findings are related both to the observed tendency in screen reading to encourage skimming, skipping, and browsing, and also to the screen's intrinsic lack of a book's concrete, spatial dimension, which tells us where things are.

There is no resolution yet of how all of this affects student understanding. Some recent studies find no significant medium-based differences in students' general comprehension, at least when a text is relatively short, while other studies, particularly by Israeli scholars, show more specific differences favoring print reading when time is taken into account. Liu raises the question of whether the length of the texts might explain the different results among the studies done to date and whether longer texts would elicit more varied performances.

What can be stated at the present time is that in the Mangen-led research, sequencing of information and memory for detail change for the worse when subjects read on a screen. Andrew Piper and David Ulin argue that the capacity to sequence matters—both in the physical world and on the printed page, even if less on digital devices. In reading as in life, Piper insists, human beings need a "sense of the pathway," a knowledge of where they are in time and space that, when needed, allows them to return to things over and over again and learn from them—something he refers to as the *technology of recurrence*.

From a very different perspective, in his thought-provoking essay "Losing Our Way in the World," the Harvard physicist John Huth writes about the more universal importance of knowing where we are in time and space and what happens when we fail to connect the details of that knowledge into a larger picture. "Sadly, we often atomize knowledge into pieces that don't have a home in a larger conceptual framework. When this happens, we surrender meaning to guardians of knowledge and it loses its personal value."

The question that arises is whether the diminution of such physical knowledge on digital mediums—the sense of being both elsewhere and nowhere on the screen—adversely affects how readers grasp the details of what they read and, at a deeper level, how they reach that close-to-palpable place where the reading act can transport us. The literary critic Michael Dirda uses this physical dimension to direct our thoughts to something far deeper within the reading experience. After comparing the reading of books on screens to staying in sterile hotel rooms, he poignantly states, "Books are *home*—real, physical things you can love and cherish." The physical realness of books contributes to our ability to enter the space where we can dwell unjudged with our hard-won thoughts and multilayered emotions and feel we have found our way home.

In this sense, physicality proffers something both psychologically and tactilely tangible. Piper, Mangen, and the literary scholar Karin Littau expand this with their emphasis on the usually unheralded role that touch plays in how we approach words and understand them in the overall text. In Piper's view, the sensory dimension of print reading adds an important redundancy to information—a kind of "geometry" to words—which contributes to our overall understanding of what we read. If you think back to Letter Two and all the things that contribute to how we process words, Piper's view makes physiological sense. The more we know about a word, the more activated our brain becomes and the more levels of meaning are available. Piper suggests that touch adds another dimension to what is activated when we read a word in print form that may go missing on the screen.

There is a very old concept called a *set* in psychological research that helps explain the less linear, less sequenced, and potentially less nuanced ways many of us are now reading, regardless of medium. When we read for hours on a screen whose characteristics involve a rapid speed of information processing, we develop an unconscious set toward reading based on how we read during most of our digital-based hours. If most of those hours involve reading on the distraction-saturated Internet, where sequential thinking is less important and less used, we begin to read that way even when we turn off the screen and pick up a book or newspaper.

There is a worrisome and potentially more lasting aspect to this "bleeding over" effect, related to the neuroplasticity concepts emphasized in these letters: the more we read digitally, the more our underlying brain circuitry reflects the characteristics of that medium. In his book *The Shallows*, Nicholas Carr reminds us of a concern raised by Stanley Kubrick that in a digital culture we should not be worrying so much about whether the computer will become like us, but whether we will become like it. Reading research buttresses the validity of such concerns. Our reading-brain circuit is the sum of many processes, most of which are continuously being shaped by the environmental demands placed on them—or not.

For example, the noted changes in the quality of our attention are intrinsically related to potential changes in memory, particularly the shorter form called *working memory*. Think back to the first spotlights that went on under the circus tent for reading. We use working memory to hold information for a short time so that we can attend to it and manipulate it for a cognitive function—for example, holding numbers "in mind" for a math problem, letters while decoding a word, or words in brief memory while reading a sentence. For many years there was an almost universally held principle that the psychologist George Miller called the "7 plus or minus 2 rule" for working memory. The "7 plus or minus 2 rule" is why most phone numbers have seven numbers, with the area code, according to Miller, able to be recalled as one unit in memory. In his later memoirs, Miller wrote that the number seven was more metaphorical than precise. In fact, the recent work on working memory suggests that the number of bits we can hold without errors may well be "4 plus or minus 1."

Until recently, I assumed that Miller's metaphorical number seven was simply an inaccurate gloss, given the newer calculations about our working memory. I have begun to question this assumption. Naomi Baron cited a 2008 report commissioned by Lloyds TSB Insurance and rather dramatically titled "'Five-Minute-Memory' Costs Brits £1.6 Billion," in which the average attention span of adults was determined to be a little over five minutes. Although five minutes may seem rather unimpressive, more noteworthy is that it is barely half of what it was only a decade earlier.

Importantly, even though the report was more about attention than about working memory, the connections between the two have been well studied. An Ariadne-like thread may well link the noted problems with remembering narratives when reading on a digital medium with changes in attention span and memory. Recall that Socrates vigorously argued that written language, touted by others as an aide-mémoire, was in practice a "recipe for forgetting." Socrates felt that if humans began to rely on the written form of language to preserve their knowledge, they would no longer utilize their highly developed memories as well as before. In our kindred transition from a literate to a digital culture, we must consider whether different forms of memory will also change with a newer "recipe."

Our culture's recipe would not be so much for forgetting, but for never remembering the same way in the first place: first because we are splitting our attention too much for our working memory to function optimally; and second, because we assume that in a digital world, we do not need to remember in the ways we remembered in the past. The current variation of Socrates' worry is that our increased reliance on external forms of memory, combined with the attention-dividing bombardment by multiple sources of information, is cumulatively altering the quality and capacities of our working memory and ultimately its consolidation in long-term memory. And indeed there are some glum estimates that indicate that the average memory span of many adults has diminished by more than 50 percent over the last decade. We will need to vigilantly replicate such studies over time. But the chain does not end there.

WHAT WE READ

Everything having to do with reading is connected: reader, author, publisher, book; in other words, the present and future of reading. Over time the effects of changing behaviors in how we are reading cannot help but influence what we read and how it is written. The implications of these changes could impact various aspects of written language, from an individual's ability to give sufficient time to unpacking the multiple layers of meaning in words; to a writer's use of words and sentences that demand and reward complex analysis; to a culture's appreciation of its writers. Italo Calvino wrote about this in a single, unalterable sentence:

> For the prose writer: success consists in felicity of verbal expression, which every so often may result from a quick flash of inspiration but as a rule involves a

patient search for the *mot juste*, for the sentence in which every word is unalterable, the most effective marriage of sound and concepts . . . concise, concentrated and memorable.

Will the word-spotting, skimming readers of the twenty-first century miss half of the words in Calvino's phrase-filled utterance? Or, as they catch "inspiration," "*mot juste*," "effective marriage," and "memorable," will they think they got the point, the gist—never realizing that they have missed the tracks of a writer's hard-won truth and the beauty in each carefully selected, deliberately sequenced word and thought? Calvino dedicated his life to attaining a form of precision, refinement, and lightness in writing that may become invisible or, worse, irrelevant to the skimming readers we could become.

Recently I read an essay on reading by the editor of *Notre Dame Magazine*, Kerry Temple, who made this observation:

When I read a manuscript sent to us for potential publication, I print it out. I make sure I read the hard copy, not the screen version. That helps me really read the words, pay closer attention, fully engage the story being told, *be* with it as I read it.

I do this because my job as an editor asks me to care about the depth and quality and nuance and substance of the stories we tell on our pages. I also do this because, as a writer, I know the labor put into crafting prose. The writer deserves my attention to detail; I honor the transaction with my thoughtful focus, by being fully present during the encounter.

This passage exemplifies what one hopes for in the encounter between the writer's intention and the reader's

attention. Less happily, however, we are beginning to observe the direct and indirect influence of the digital word-
spotting, text-grazing reading patterns of contemporary
readers—how things are read—on how texts are being
written. When publishers are forced to consider the needs
of a different reader, one whose typical skimming style is ill
suited to long, densely worded texts, to complex thoughts
not easily (or quickly) grasped, or to words deemed less
than necessary, the culture suffers in ways we cannot measure. Things go missing in such a context, unnoted till they
are absent.

Not too long ago, David Brooks wrote a column on
beauty and his quiet feeling that unannounced . . . it has
gone missing. Brooks placed no blame. He provided no an-
odyne solution. He simply looked at what is being imperceptibly lost by us as we "accidentally" abandon a view of
the world in which beauty, truth, and goodness are inextri-
cably linked and where the very perception of beauty can
be a path to a life where virtue and nobility have rightful
place.

Like insight, the perception of beauty, whether in reading or in art, emerges out of many of the same capacities
that compose deep reading. And, like insight, only the
time we give to those capacities allows our perception of
beauty to "father forth" long enough for us to see, recognize, and understand more. For just as reading is not solely
visual, beauty is not simply about the senses. In her essay
"Decline," Marilynne Robinson wrote that beauty among
other important things is a "strategy of emphasis. If it is
not recognized, the text is not understood." Beauty helps
us attend to what is most important. If our perception of
beauty becomes reduced to skimming like a water strider
across the thin surface of words, we will miss the depths

below; we will never be led by beauty to learn and understand what lies beneath.

Before the transition into our present digital culture,
Calvino gave our millennium a related and prescient set
of insights into some of the far-reaching ramifications of
these issues:

> In an age when other fantastically speedy, wide-
> spread media are triumphing, and running the risk
> of flattening all communication onto a single, homo-
> geneous surface, the function of literature is com-
> munication between things that are different simply
> because they are different, not blunting but even
> sharpening the differences between them, following
> the true bent of written language.

Calvino, who devoted his whole life to translating difficult thoughts into words, left us with the plea that language
in all its complexity not be "flattened" by us. The future of
language is linked both to the sustained efforts by writers
to find those words that direct us to their hard-won thought
and to the sustained efforts by readers to reciprocate by applying their best thought to what is read. I worry that we
are one quick step removed from recognizing the beauty
in what is written. I worry that we are even closer to the
stripping away of complex thoughts when they do not fit the
memory-enfeebling restriction on the number of characters
used to convey them. Or when they are buried in the last,
least read, twentieth page of a Google search. The digital
chain that leads from the proliferation of information to
the gruel-thin, eye-byte servings consumed daily by many
of us will need more than societal vigilance, lest the quality
of our attention and memory, the perception of beauty and

recognition of truth, and the complex decision-making capacities based on all of these atrophy along the way.

When language and thought atrophy, when complexity wanes and everything becomes more and more the same, we run great risks in society politic—whether from extremists in a religion or a political organization or, less obviously, from advertisers. Whether cruelly enforced or subtly reinforced, homogenization in groups, societies, or language can lead to the elimination of whatever is different or "other." The protection of diversity within human society is a principle that was embodied in our Constitution and long before that in our genetic cerebrodiversity. As described by geneticists, futurists, and most recently Toni Morrison in her book *The Origin of Others*, diversity enhances the advancement of our species' development, the quality of our life on our connected planet, and even our survival.

Within this overarching context, we must work to protect and preserve the rich, expansive, unflattened uses of language. When nurtured, human language provides the most perfect vehicle for the creation of uncircumscribed, never-before-imagined thoughts, which in turn provide the basis for advances in our collective intelligence. The converse is also true, with insidious implications for every one of us.

Not long ago, I discussed these admittedly dark and heavy thoughts in the lightest of settings: a summer's conversation on a walk in the French Alps with the Italian publisher Dr. Aurelio Maria Mottola. As we walked higher and higher to where the trees began to thin, I told him of my worries about the possible effects of the culture's trend toward language homogenization: from the narrowing of an author's word choices to briefer manuscripts to a more constrained use of syntactic complexity and figurative language, both of which require background knowledge that can no longer be assumed.

What, then, will be the fate, he asked, of books and poems filled with metaphors and analogies whose referents are
no longer shared knowledge? What would happen if a culture's shared repertoire of allusions—metaphors from the
Bible, myths, and fables; remembered lines of poems; characters from stories—begins to shrink and gradually disappear? What will happen, this learned publisher, who reads
in multiple languages, asked, if the "language of books" no
longer fits the culture's cognitive style—fast, heavily visual,
and artificially truncated? Will writing change and with it
the reader, the writer, the publisher, language itself? Are
we each witnessing in our different professions the beginning of a retreat from more intellectually demanding forms
of language until—like the ill-fated procrustean bed—it
conforms to the imperceptibly narrowing norms of reading
on ever smaller screens?

We stopped somewhere in the beautiful landscape and
tried to rescue our walk from the unwelcome direction
of our thoughts. Isn't the nature of language to expand
and change each epoch? we asked each other. Isn't the
very history of writing our own epoch's best reassurance?
Doesn't the plasticity of the reading brain provide the ideal
mechanism to accommodate diverse modes of reading and
writing?

We must not lose what we have gained, I said softly, as
much to you, dear readers, as to my then silent companion
on that summer walk. Some of you, no doubt, will think
that I protest too much, and that only the elite parts of any
population will miss the shelves of older books and poems
that pass out of favor with clockwork regularity age after
age, generation after generation. But it is the very opposite
of elitism that propels my worries. I write this book and
conduct my research today only because the dedication of
my parents and of a few deeply committed teachers from

the School Sisters of Notre Dame in a two-room, eight-grade schoolhouse gave me reason as a child to read the "great literature" of the past. Only those books prepared me not to leave the coal miners and farmers in my tiny midwestern town but to understand each of those still dear people and the world outside Eldorado, Illinois, in whole new ways. Words, stories, books allowed me to have not so much a quiet eye—never, perhaps, my forte when I was young—but a widened gaze at worlds I could never have imagined from my very small vantage point over Walnut Street, where I met Emily Dickinson, Charlotte Brontë, and Margaret Mitchell for the first time. As Alberto Manguel remarked about his own similarly book-built knowledge store, "Everything proceeds in geometric progression based on what is known and what is remembered every time we read something new."

There is no question that today's children and youth will leave their own version of Walnut Street to discover untold worlds through the Internet with all its amazing potential to connect people and ideas from around the globe. But before they do and as they do, I want them to be actively building their own uniquely formed internal bases of knowledge that embody what they learn from both the books off those shelves and graphic novels by Gene Luen Yang and Mark Danielewski. I want them to learn to read and to remember, for this is the foundation for who they will become and how they will think and decide the shape of their future and ours.

I have taught hundreds of bright, well-educated undergraduates over the last years. I feel daily uplifted by their intelligence and their desire to contribute something of value to our world, a particular goal of the university where I have done my work. But the reality is that more and more of them are wonderfully adept at programming languages, but have a harder and harder time when I refer to a "coat of many colors," "the quality of mercy," or "the road not taken"—and this in New England. The question I must confront in my self-appointed role as reading worrier is whether the carefully built internal platforms are being sufficiently formed in our young before they automatically turn to their default intelligence and look up an unknown name or concept. It is not that I prefer internal to external platforms of knowledge; I want both, but the internal one has to be sufficiently formed before automatic reliance on the external ones takes over. Only in this developmental sequence do I trust that they will know when they do not know.

The issue, therefore, is never just about how many words we consume or even how we read in the digital culture. It is about the significant effects of how much we read upon how we read and the effects of both upon what we read and remember. It does not end, however, with what we read, but rather continues on, as what we read further changes the next link in the chain, how things are written.

HOW THINGS ARE WRITTEN

During my graduate courses with Carol and Noam Chomsky, my view of language shifted from my former emphasis on the beauty of the word to the study of the word within the structures of language. That shift impressed upon me what my previous study of literature had not: that the varied processes of language, particularly syntax, reflect the convolutions of our thoughts. As the Russian psychologist Lev Vygotsky wrote in a remarkable book called *Thought and Language*, written language not only reflects our most difficult thoughts, it propels them further.

Within the context of written language's influence on intellectual development, consider the increasing unease of many English professors in universities and in high schools that growing numbers of their students no longer have the "patience" to read literature from the nineteenth century and early half of the twentieth century. If we think about Herman Melville's *Moby-Dick* and George Eliot's *Middlemarch*, two of the finest examples of nineteenth-century literature in the English language, the density of the sentences in these books and the cognitive analyses required by the reader to understand them are substantial. One of my favorite sentences in *Middlemarch* illustrates this as it describes the moment of insight when poor Dorothea discovers the limits of her older husband's assumed genius—on her honeymoon!

How was it that in the weeks since her marriage Dorothea had not distinctly observed but felt with a stifling depression that the large vistas and wide fresh air which she had dreamed of finding in her husband's mind were replaced by anterooms and winding passages which seemed to lead nowhither?

There is, to be sure, no shortage of words, phrases, and clauses here. Yet Eliot's dense grammar and the "winding" structure of the sentence are a near-perfect simulation of the aimless meanderings of Mr. Casaubon's mind, which also wandered "nowhither." It is fair to say that a generation of young people raised on the Internet and Twitter, simultaneously flooded with volumes of words and accustomed to expending only 140 characters to write their thoughts, would have a difficult time appreciating this sentence or reading Melville or Eliot, much less the 150 to 300-plus words in a typical sentence by Proust.

To be sure, some of the issues here have to do with the expected changes in language usage that occur from one historical epoch to the next, just as Dr. Mottola expressed. Holding this thought in mind, in the most unscientific of exercises I went to my bookshelves and pulled down three recent bestselling novels by well-known, respected authors and three early-twentieth-century novels. I had enjoyed each of the books tremendously, and I simply wanted to see what a random examination of grammatical structures in the contemporary works might reveal, if anything. I used a very simplified (that is, nonscientific) version of what reading researchers such as my former mentor Jeanne Chall call *readability formulae*, which evaluate the age-level appropriateness of different texts. (I will admit that I assiduously, successfully avoided studying such formulae during my entire graduate program.) I randomly scanned pages from each of the books to calculate the average number of words per sentence and the average number of phrases and clauses per sentence and paragraph. Despite considerable differences in style and content, the average sentence length across the current bestselling novels appeared to be less than half the average counts from those in the early- to mid-twentieth-century works, with dramatically fewer clauses and phrases per sentence.

The trend away from density in prose does not need formal readability formulae for us to observe it in daily life. The question is whether we are observing an accelerated realignment between reading style (how we read) and writing style (what we read), and if so, does it matter? My superficial sampling prohibits any facile judgments: either about changes from one epoch's style of writing to another; or about whether the noted changes reflect the characteristics of the dominant medium, or, more ominously, the complexity of thought embodied in the works. It would be a gross error to suggest that the depth of an author's thoughts is directly correlated to the syntactic density of a

work. I have often written that we can all appreciate both
Hemingway and George Eliot. Nevertheless, I have begun
to question the cognitive loss of not being willing or, perhaps in the future, even able to navigate the demands of the
complex concepts in denser prose. I am increasingly worried, therefore, about the relationship between the number
of characters with which we choose to read or write and
how we think. Never more so than now; never more so
than with our young adults, or with those who would lead
our governments around the world.

TL; DR (Too long; didn't read). The critical relationship
between the quality of reading and the quality of thought
is influenced heavily by changes in attention and what I
have called, more intuitively than scientifically, *cognitive
patience.* Some of the more disconcerting and surprising
letters I have received over the last years have come from
professors of literature and the social sciences who are
flummoxed over their college students' impatience with
older, denser American literature and writing. One chair
of a well-known English department wrote that he could
no longer teach his once sought-after seminar on Henry
James because too few students today wanted or were able
to read James. Among those professors, the most frequent
observations are two. The first is that students have become increasingly less patient with the time it takes to understand the syntactically demanding sentence structures
in denser texts and increasingly averse to the effort needed
to go deeper into their analysis.

The second is that student writing is deteriorating. I
have, to be sure, heard this criticism of undergraduates as
long as I have been teaching. The question is nevertheless
important for every age to confront. In our epoch, we must
ask whether current students' diminishing familiarity with
conceptually demanding prose and the daily truncating of

their writing on social media is affecting their writing in more negative ways than in the past. There are two issues that may also be related to cognitive patience that are at work in student writing. In a project designed to track the use of citations by students, most student citations referred either to the first page of the source they cited or to the last three pages.

One can only wonder whether the pages in between the first and last were ever read, or whether the entire article was read with the *F* or zigzag style Liu describes: that is, reading the first page, a little in the middle, and then the last pages. If so, the background knowledge, argumentation, and supporting evidence found in the body of most resources has gone either skimmed or largely unread. Such a mode of reading will eventually find its way into less well constructed and less persuasively supported writing by students who are conceptually skimming in both their reading and their writing.

Several professors went on in their letters to admit with no small ambivalence that they were now assigning collections of short stories to deal with the shorter attention spans of their students. The intrinsic value of the short-story genre is not in question. But just as the reported decline of empathy in our young people requires our collective scrutiny and understanding, so also does the more and more frequent observation that students are shying away from longer, more difficult texts and writing less well than in the immediate past. The central issue is not their intelligence, nor, more than likely, even their lack of familiarity with different styles of writing. Rather, it may come back to a lack of cognitive patience with demanding critical analytic thinking and a concomitant failure to acquire the cognitive persistence, what the psychologist Angela Duckworth famously called "grit," nurtured by the very genres being avoided. Just as earlier I described how a lack of background knowledge and critical analytical skills can render any reader susceptible to unadjudicated or even false information, the insufficient formation and lack of use of these complex intellectual skills can render our young people less able to read and write well and therefore less prepared for their own futures.

These are the very intellectual skills and personal attributes that provide young adults with the most important foundation for being able to recognize and manage the unavoidable changes and complexities ahead of them. Their development in the college years prepares them for the far more challenging forms of intellectual tenacity required of them after graduation: whether it is to write well-argued reports, documents, and briefs in their future professional lives; to critically read and evaluate the worth of a referendum, a court decision, medical documents, wills, investigative journalism, or a political candidate's personal record; or even to differentiate truth from falsehood in the escalating issues around false news and reports. A democratic society requires the careful development of these abilities in its citizens, both young and old.

In "Internet of Stings," Jennifer Howard began one of the more disconcerting essays about some of these issues that came up in interviews with one of the purveyors of false news:

> As one master of the fake-news genre told the *Washington Post*: "Honestly, people are definitely dumber. They just keep passing stuff around. Nobody fact-checks anything anymore." Separating truth from fiction takes time, information literacy, and an open mind, all of which seem in short supply in a distracted, polarized culture. We love to share instantly—and that makes us easy to manipulate.

There are many tough issues here for students, teachers, parents, and the members of our republic. How our citizens think, decide, and vote depends on their collective ability to navigate the complex realities of a digital milieu with intellects not just capable of, but accustomed to higher-level understanding and analysis. It is no longer only a matter of which medium is better for what; it is a question of how the optimal mode of thought in our children and our young adults and ourselves can be fostered in this moment of history.

These are hardly new thoughts either for me or for others. Both Marshall McLuhan's iconic messages about the medium's influences on us and Walter Ong's more philosophical exhortations harken back once again to Socrates' original worry that reading would permanently change thinking. "If men learn this, it will implant forgetfulness in their souls; they will cease to exercise memory because they rely on that which is written, calling things to remembrance no longer from within themselves, but by means of external marks." Certainly Socrates never had time to understand the potential value of having both internal and external sources of memory, but we do. Yet we don't take the time we have to attend to the full import of the changes in how we read and think upon our daily lives.

The Jesuit scholar Walter Ong helped situate the spot-on accuracy of some of Socrates' concerns and also their shortcomings when applied to contemporary society. He argued that our intellectual evolution is not so much about how one medium of communication differs from another but rather what happens to human beings who are steeped in both. From Ong's perspective, what will our age's readers—who inherit both literacy-based and digital cultures—become? Are the changes in oral language, reading, and writing so subtle that before we attend to them, we will have forgotten what we thought to be true, fine, virtuous, and essential to human thought? Or can we use the sum of the present knowledge and our inferences based on it to select what is best from both mediums and teach this to our young?

The will that is necessary to answer these questions begins with a deeper examination of our own reading lives, begun in the last letters. Do you, my reader, read with less attention and perhaps even less memory for what you have read? Do you notice when reading on a screen that you are increasingly reading for key words and skimming over the rest? Has this habit or style of screen reading bled over to your reading of hard copy? Do you find yourself reading the same passage over and over to understand its meaning? Do you suspect when you write that your ability to express the crux of your thoughts is subtly slipping or diminished? Have you become so inured to quick précis of information that you no longer feel the need or possess the time for your own analyses of this information? Do you find yourself gradually avoiding denser, more complex analyses, even those that are readily available? Very important, are you less able to find the same enveloping pleasure you once derived from your former reading self? Have you, in fact, begun to suspect that you no longer have the cerebral patience to plow through a long and demanding article or book? What if, one day, you pause and wonder if yourself are truly changing and, worst of all, do not have the time to do a thing about it?

A Case Study of the Digital Chain Hypothesis

And so I come to the story of my unsettling. Hardly the grist of a bestseller, the plotline is this: researcher of reading and its changes in a digital culture awakens one day and is forced to confront whether she has changed, too. It remains for me a pitiful tale with a few difficult lessons for both of us and words to eat for me.

Calvino once wrote that Washington Irving's "Rip Van Winkle" has "acquired the status of a foundation myth for your ever-changing society." It is certainly the case for me, and twice now. As I described in my first letter, my initial "awakening" experience happened at the end of writing *Proust and the Squid.* After seven years of researching the reading brain, I looked around and realized that my entire subject matter had changed. Reading was no longer the entity it was when I had begun.

The second experience hit still closer to home. Despite all my research on the changing reading brain, I never realized that the same things were true for me until the effects were close to the proverbial done deal. They began innocently enough. Like everyone else, with increasing responsibilities in my professional and personal lives and an ever-increasing load of what I had to read and write in any given day on any number of digital mediums, I began to make little compromises. I still tried to use email more like a note in an envelope—a social greeting with its own forms of politeness. But every note was becoming shorter and terser. There was no waiting for the perfect moment to write tranquil thoughts, the admittedly elusive goal of my former style. I just did my best at any given moment and hoped for cosmic forgiveness for failing to meet the projected expectations of the recipients of all my communications.

As for reading, I increasingly depended on Google, Google Scholar, daily/weekly summaries in journals such as *Science,* online news, online *New Yorker* stories, and so on, for what I thought I needed to know or needed to read later in more depth. Various subscriptions to newspapers and magazines came and went. I could no longer keep up with the ones that mattered most—the ones that provided the most in-depth commentaries on public life—and so . . . I didn't. I pretended to myself that I would catch up on the weekends, but the unmet deadlines of the week bled into the weekends, whose earlier restorative goals gradually disappeared.

The next vanishing act was the heretofore highly anticipated books that always sat at my bedside waiting to be read. In their place in the final minutes of my day came the last emails, so that I could sleep feeling "virtuous," instead of comforted by a reflection of Marcus Aurelius or stilled by reading the books of Kent Haruf or Wendell Berry, in which little happens, save for the recollected insights of people who are guided by the earth's rhythms, human love, and tested virtue and whose observations quiet the unquiet mind and restless heart.

I still bought many books, but more and more I read *in* them, rather than being whisked away *by* them. At some time impossible to pinpoint, I had begun to read more to be informed than to be immersed, much less to be transported.

With that unwelcome realization, I stopped in my own version of suspended disbelief: Was it possible that I had become the reader for whom I was losing my weekends writing about and indeed for? Hubris alone prevented my acceptance of such a scenario. Rather, like any scientist confronting a researchable question, I set up an experiment. Unlike every other study I conducted, I was the single subject in a single-cell design. My null hypothesis, if you will, was that I had not changed my reading style; rather, only the time I had available for reading had changed. I could prove that simply enough by controlling for it by setting aside the same amount of time every day and faithfully observing my own reading of a linguistically difficult, conceptually demanding novel, one that had been one of my favorite books when I was younger. I would know the plot. There would be no suspense or mystery involved. I would have only to analyze what I was doing during my reading in the same way that I might analyze what a person with dyslexia does when he or she is reading in my research center.

With little hesitation I chose Hermann Hesse's *Magister Ludi*, also known as *The Glass Bead Game*, which was cited when Hesse received the Nobel Prize in Literature in 1946. To say I began the experiment with the most cheerful of dispositions is no exaggeration. I was practically gleeful at the idea that I would be forcing myself to reread one of the most influential books of my earlier years.

Force became the operative word. When I began to read *Magister Ludi*, I experienced the literary equivalent of a punch to the cortex. I could not read it. The style seemed obdurately opaque to me: too dense (!) with unnecessarily difficult words and sentences whose snakelike constructions obfuscated, rather than illuminated, meaning for me. The pace of action was impossible. A bunch of monks slowly walking up and down stairs was the only image that came to mind. It was as if someone had poured thick molasses over my brain whenever I picked *Magister Ludi* up to read.

To compensate, first I consciously tried to read the text less quickly, to no avail. The rapid speed to which I had become accustomed while reading my daily gigabytes of material did not allow me to slow down enough to grasp whatever Hesse was conveying. I did not need a galvanic skin response test to know that my skin was sweating faintly. I was breathing more heavily, and my pulse rate was probably elevated. I would not have wanted to know my cortisol levels. I hated the book. I hated the whole so-called experiment, which was anything but scientific in the first place. Finally, I wondered how on earth I had ever thought that this was one of the great twentieth-century novels, Hesse's Nobel Prize notwithstanding. It was a different time. It would never be well received now. Hesse could probably not have found a publisher for the book today.

Case closed, I thought as I shoved *Magister Ludi* unceremoniously back between Hemingway and Hesse's much less demanding *Siddhartha* on my neatly alphabetized bookshelf, the one filled with the books that had largely formed who I am and how I think. It scarcely mattered that I had failed my own test. No one would care or know but me. No one would be the wiser.

As for any wisdom on my own part, the inescapable conclusion—which I had no intention of sharing with anyone else—was that I had changed in ways I would never have predicted. I now read on the surface and very quickly; in fact, I read too fast to comprehend deeper levels, which forced me constantly to go back and reread the same sentence over and over with increasing frustration; I was impatient with the number of clauses and phrases per sentence, as if I had never reverentially encountered the much longer sentences of Proust and Thomas Mann; I was completely offended by the number of words Hesse felt necessary to use in every other sentence; and finally, my so-called deep-reading processes never "surfaced." There. I had changed. I was Ionesco's rhinoceros, too. "So what?" I muttered aloud to no one in particular.

The experiment was a disaster. It would have gone no further than the privacy of my bookshelves had it not been for two quietly disturbing thoughts: First, the bookshelves were filled with my friends—including Hermann Hesse—

whose collective, formative influences upon me had been second only to those of my family and my teachers. Was I, for all purposes, about to desert the friends of a lifetime, relegating most of them perfunctorily to their alphabetized place, shelved in a different time? Second, over the years I had told a thousand children with dyslexia that failures, like enemies, can be our best teachers, if we can see them as opportunities for recognizing what we need to change. With the reading equivalent of "gritting one's teeth," I forced myself to return to task, but this time for mercifully brief, concentrated, twenty-minute intervals. I was cagily vague with myself about how many days I would give this unplanned, unpleasant, and undesired Phase Two of the experiment.

It took two weeks. Somewhere near the end of that many days, I experienced a much less dramatic form of St. Paul of Tarsus's epiphany. No flash of light or brilliant insight. I simply felt, at last, that I was home again, returned to my former reading self. The pace of my reading now matched the pace of action in the book. I slowed down or speeded up with it. I no longer imposed upon Hesse's words and clause-lined sentences either the speed or the spasmodic quality of attention that I had unconsciously grown accustomed to in my online reading style.

In her wonderful book *Rereadings*, Anne Fadiman compared reading to rereading a book: "the former had more velocity; the latter had more depth." My experience as a digital screen reader trying to reread Hesse's masterpiece was the opposite: I had tried to reread it as quickly as possible, and I had failed. Indeed, Naomi Baron predicted that the shift to screen reading would diminish our desire for rereading, which would be a great loss since each age at which we read brings a different person to the text. In my case, only when I forced myself to enter the book did I experience, first, slowing down; second, becoming immersed in the other world in the book; and third, being lifted out of my own. During the process, my world slowed down—just a little—as I recovered my lost way of reading.

As my small experiment showed me, my own reading circuit had adapted to the demands put upon it, and though I had depressingly little consciousness of this, my reading behaviors (or style) had changed along the way. In other words, my grafted, spasmodic, online style, while appropriate for much of my day's ordinary reading, had been transferred indiscriminately to all of my reading, rendering my former immersion in more difficult texts less and less satisfying. I did not go further and test my comprehension for possible changes. I will admit that I did not want to know that. I simply wanted to regain what I had almost lost.

Ultimately my simplistic experiment was a way to confront issues that are crucial for each of us who is steeped in both print and digital mediums. In Ong's terms, the question I confronted involved a recognition of the ways I had become transformed by the two different modes of reading. Perhaps equally important, given the reality of my daily straddling of both forms of communication, the filial question was Klinkenborg's: What would now become of the reader I had been?

There is a very simple, very beautiful Native American story I have always remembered. In this story a grandfather is telling his young grandson about life. He tells the little boy that in every person there are two wolves, who live in one's breast and who are always at war with each other. The first wolf is very aggressive and full of violence and hate toward the world. The second wolf is peaceful and full of light and love. The little boy anxiously asks his grandfather which wolf wins. The grandfather replies, "The one you feed."

THE LAST LINK IN THE DIGITAL CHAIN: WHY WE READ

It is within the context of feeding the "second wolf" that I tell you the real denouement of my experiment in rereading Hesse: I read *Magister Ludi* a third time. Not for any experimental reason, simply for the peace I felt in returning to my former reading life. The novelist Allegra Goodman wrote something wonderful about the process of unfolding that occurs in rereading a beloved book: "Like pleated fabric, the text reveals different parts . . . at different times. And yet every time the text unfolds, . . . the reader adds new wrinkles. Memory and experience press themselves into each reading so that each encounter informs the next." Each time I reread the book, I remembered something essential about the person I had thought I was when I had first read Hesse. By the end I had recaptured both why I had read the book with such joy then and, ironically perhaps, what reading had meant to me before I had become a reading researcher.

There may well be as many reasons why we read as there are readers. But the very raising to consciousness of the question *why we read* has elicited some of the most thought-provoking responses by some of the world's most beloved writers. I would ask you to raise it for yourself before more time passes. After I refound my former reading self, the answer that came to me is this: I read both to find fresh reason to love this world and also to leave this world behind—to enter a space where I can glimpse what lies beyond my imagination, outside my knowledge and my experience of life, and sometimes, like the poet Federico García Lorca, where I can "go very far, to give me back my ancient soul of a child."

This harkens to something else that Hesse wrote in a little-known essay called "The Magic of the Book":

Among the many worlds which man did not receive
as a gift of nature, but which he created with his own
spirit, the world of books is the greatest. Every child,
scrawling his first letters on his slate and attempting
to read for the first time, in so doing, enters an artificial and most complicated world: to know the laws
and rules of this world completely and to practice
them perfectly, no single human life is long enough.
Without words, without writing, and without books
there would be no history, there could be no concept
of humanity.

Hesse's "many worlds" and Lorca's dream of recovering
the "ancient soul of a child" are the best ways I have of
introducing you to the next Letter, about the children who
come after us and the unique legacy of the reading life we
hope to give them and their children and their children's
children.

Sincerely yours,
Your Author

THE RAISING OF CHILDREN IN A DIGITAL AGE

Children are a sign. They are a sign of hope, a sign of life, but also a *"diagnostic" sign*, a marker indicating the health of families, society, and the entire world. Wherever children are accepted, loved, cared for and protected, the family is healthy, society is more healthy and the world is more human.

—Pope Francis

Every medium has its strengths and weaknesses; every medium develops some cognitive skills at the expense of others. Although . . . the Internet may develop impressive visual intelligence, the cost seems to be deep processing: mindful knowledge acquisition, inductive analysis, critical thinking, imagination, and reflection.

—Patricia Greenfield

Dear Reader,

Once when my children were young, they asked me to tell them one more time what I do when I go to work. We had just returned from a visit to one set of their grandparents, who live in the heart of the Midwest. There the children saw fields of corn and beans and herds of cattle and horses that captured their city-bred imaginations. Spontaneously, I found myself saying, "I am a farmer of children!"

They laughed and found that a very fine answer, much better than a teacher or a researcher of the reading brain. I liked that answer, too, and I found myself storing it away as my private way of looking at what I do with my life.

I recall it now because that is what this letter is all about: how we "raise" children, children who are the inheritors of the twentieth century and the progenitors of the twenty-first. They are "Mine own, and not mine own," as Shakespeare described one of the many forms of love, in *A Midsummer Night's Dream*. They are our own, and not our own. Further, they are on the verge of becoming more different from us—their parents and grandparents and great-grandparents—than at any time since the last other, great transitions in modes of communication: the time between Socrates' oral culture and Aristotle's written one and the period following Gutenberg.

There will always be either a chasm or just a little gully of differences between parents and their children across every epoch. I am less interested in the degree of difference between our digitally raised children and ourselves than in an understanding of what is best for children's development regardless of milieu and, in particular, within this exponentially changing milieu. There is no going back, and with some historical digressions aside, there almost never has been. That accepted reality, however, must not deter anyone from informed, compassionate, critical analyses of who we have been, who we are, and the changes that are quietly shaping our children every day.

The changes are many and various. The tough questions raised in the previous letters come to roost like chickens on a fence in the raising of our children. They require of us a developmental version of the issues summarized to this point: Will the time-consuming, cognitively demanding deep-reading processes atrophy or be gradually lost within a culture whose principal mediums advantage speed, immediacy, high levels of stimulation, multitasking, and large amounts of information?

Loss, however, in this question implies the existence of a well-formed, fully elaborated circuitry. The reality is that each new reader—that is, each child—must build a wholly new reading circuit. Our children can form a very simple circuit for learning to read and acquire a basic level of decoding, or they can go on to develop highly elaborated reading circuits that add more and more sophisticated intellectual processes over time. There will be many differences in how the circuit develops along the way, based on individual children's characteristics, the type of reading instruction and support they receive, and, critical to our discussion, the medium(s) in which they are reading. The medium's characteristics or affordances—from physicality to attention-capturing options—add a new, much less understood dimension to influences on the reading circuit's development. As the UCLA psychologist Patricia Greenfield demonstrates in her work, the basic, commonsense principle is that the more exposure to (time spent with) any medium, the more the characteristics of the medium (affordances) will influence the characteristics of the viewer (learner). The medium is the messenger to the cortex, and it begins to shape it from the very start.

The not-yet-formed reading circuits of the young, therefore, present us with unique challenges and a complex set of questions: First, will the early-developing cognitive components of the reading circuit be altered by digital media before, while, and after children learn to read? In particular, what will happen to the development of their attention, memory, and background knowledge—processes known to be affected in adults by multitasking, rapidity, and distraction? Second, if they are affected, will such changes alter the makeup of the resulting expert reading circuit and/or the motivation to form and sustain deep-reading capacities? Finally, what can we do to address the potential negative effects of varied digital media on reading without losing their immensely positive contributions to children and to society?

Attention and Memory in the Age of Distraction

ATTENTION

What we attend to and how we attend make all the difference in how we think. In the development of cognition, for example, children learn to focus their attention with ever more concentration and duration from infancy through adolescence. Learning to concentrate is an essential but ever more difficult challenge in a culture where distraction is omnipresent. Young adults may learn to be less affected when moving from one stimulus to another because they have more fully formed inhibitory systems that, at least in principle, provide the option of overriding continuous distraction. Not so with younger children, whose inhibitory systems and the other executive planning functions in their frontal cortex need a long time to develop. Attention, in the very young, is up for grabs.

And the digital world grabs it. In a 2015 RAND report, the average amount of time spent by three- to five-year-old children on digital devices was four hours a day, with 75 percent of children from zero to eight years old having access to digital devices, a figure that was up from 52 percent only two years earlier. In adults the use of digital devices increased by 117 percent in one year. Although the questions for society about the effects of continuous stimulation and nonstop distraction apply to us all, those effects are most urgent to understand for the young.

The psychologist Howard Gardner used the MIT scholar Seymour Papert's famous description of the child's "grasshopper mind" to describe the spasmodic way our digital young now typically "hop from point to point, distracted from the original task." Like Frank Schirrmacher, the neuroscientist Daniel Levitin places such attention-flitting, task-switching behavior within the context of our evolutionary reflex, the *novelty bias* that pulls our attention immediately toward anything new: "Humans will work just as hard to obtain a novel experience as we will to get a meal or a mate. . . . In multitasking, we unknowingly enter an addiction loop as the brain's novelty centers become rewarded for processing shiny new stimuli, to the detriment of our prefrontal cortex, which wants to stay on task and gain the rewards of sustained effort and attention. We need to train ourselves to go for the long reward, and forgo the short one."

Levitin wrote that passage in a book largely written for adult executives. His worthy lessons for adults, however, are magnified when considering young children. The child's prefrontal cortex and the entire underlying central executive system have not yet learned the "rewards of sustained effort and attention," much less the planning and inhibition that would allow a child to "forgo the short one." In other words, switching between sources of attention for the child's brain makes the perfect biological-cultural storm for adults look like a gentle downpour. With little prefrontal development on their side, children are completely at the mercy of one distraction after another, and they quickly jump from one "shiny new stimulus" to another.

Levitin claims that children can become so chronically accustomed to a continuous stream of competitors for their attention that their brains are for all purposes being bathed in hormones such as cortisol and adrenaline, the hormones more commonly associated with fight, flight, and stress. They are only three years old, or four, or sometimes even two and younger—but they are first passively receiving and then, ever so gradually, actively requiring the levels of stimulation of much older children on a regular basis. As Levitin discusses, when children and youth are surrounded with this constant level of novel, sensory stimulation, they are being projected into a continuously hyperattentive state. He suggests that "multitasking creates a dopamine-addiction feedback loop, effectively rewarding the brain for losing focus and for constantly searching for external stimulation."

It is this heightened state that may produce several relatively new phenomena in childhood today. As the clinical psychologist Catherine Steiner-Adair, the author of *The Big Disconnect: Protecting Childhood and Family Relationships in the Digital Age*, observes, the most commonly heard complaint when children are asked to go off-line is "I'm bored." Confronted with the dazzling possibilities for their attention on a nearby screen, young children quickly become awash with, then accustomed to, and ever so gradually semi-addicted to continuous sensory stimulation. When the constant level of stimulation is taken away, the children respond predictably with a seemingly overwhelming state of boredom.

"I'm Bored." There are different kinds of boredom. There is a natural boredom that is part of the woof of childhood that can often provide children with the impetus to create their own forms of entertainment and just plain fun. This is the boredom that Walter Benjamin described years ago as the "dream bird that hatches the egg of experience." But there may also be an unnatural, culturally induced, new form of boredom that follows too much digital stimulation.

This form of boredom may de-animate children in such a fashion as to prevent them from wanting to explore and create real-world experiences for themselves, particularly outside their rooms, houses, and schools. As Steiner-Adair wrote, "If they become addicted to playing on screens, children will not know how to move through that fugue state they call boredom, which is often a necessary prelude to creativity." It would be an intellectual shame to think that in the spirit of giving our children as much as we can through the many creative offerings of the latest, enhanced e-books and technological innovations, we may inadvertently deprive them of the motivation and time necessary to build their own images of what is read and to construct their own imaginative off-line worlds that are the invisible habitats of childhood.

Such cautions are neither a matter of nostalgic lament nor an exclusion of the powerful, exciting uses of the child's imagination fostered by technology. We will return to such uses a little later. Nor should worries over a "lost childhood" be dismissed as a cultural (read Western) luxury. What of the *real* lost childhoods? one might ask, in which the daily struggle to survive trumps everything else? Those children are never far from my thoughts or my work every day of my life.

But I worry about every child. Therefore, I worry very much about the cognitive-developmental trajectories of children who are so constantly stimulated and virtually entertained that they rarely want to go off (screen) to discover their own ability to entertain themselves with their own created hideaways, preferably outside where tangled bushes and sticks become "Martian land," where a tablecloth over a low-hanging tree branch becomes an Iroquois tent, where their imaginations become immersed in what they are doing, and dinner is served too soon. Time stops in places like these, and thinking lengthens. And, as the neuroscientist Fogassi memorably argues, the child's motor cortex enhances cognition and needs considerable activation, too!

These issues become more intensified for older children, as the hours spent in front of screens double and triple to twelve-plus hours a day among many adolescents, along with the level and variety of addicting enticements by digital distractions. Steiner-Adair minces no words about the addictive aspect of children's digital immersion: "Talk of addiction is not hyperbole; it is a clinical reality. . . . As adults we may choose to mess with our minds and gamble with our own neurology, but I have never met a caring parent who would knowingly risk his or her child's future in this way. And yet, we are handing these devices—that we use the language of addiction to describe—over to our children, who are even more vulnerable to . . . the impact of everyday use on their developing brains. . . . In our enthusiasm to be early adapters, and to give our kids every advantage, are we putting our children in harm's way?"

There may be no more painfully realistic description of the overwhelmingly addictive influence of digital worlds on our youth than Allegra Goodman's portrait in her new novel *The Chalk Artist* of Aidan, a highly intelligent, impressionable adolescent boy who lives both in Cambridge, Massachusetts, and in the virtual world of EverWhen. This kind and sensitive boy spends every waking hour (and most of those in which he should be sleeping) in a bloodcurdling virtual world that eventually he simply prefers, with tragic consequences. Some others, such as the psychiatrist Edward Hallowell, go so far as to suggest that we are creating a cadre of children with environmentally induced attentional deficits because of the incessant, obsession-promoting hold that digital distractions pose for a child. This clinician's concern is that the increasing number of children diagnosed with attention-based learning deficits may reflect not only better, earlier diagnoses, but also the creation of new forms of attention deficit in a generation of distracted children.

The Stanford neuroscientist Russell Poldrack and his team have investigated this question for well over a decade, including looking at physiological differences in children with and without attention-deficit diagnoses, and very recently in multitasking performances for students raised with digital media. Perhaps predictably, for children with attention deficits, there were significant differences in their prefrontal inhibitory systems, which are essential in the mental switching requirements involved in multitasking. Specifically, children with diagnosed attentional issues appeared less able to focus their attention on one task because they could not stop paying attention to all the other tasks. Given the increasing numbers of distractions that populate the digital worlds of many children, we have to ask whether greater numbers of children who are otherwise typical are becoming prone to behaviors similar to children diagnosed with attention-deficit disorder because of their environments. If so, what other effects might such changes have on different aspects of their development?

For example, a positive aspect is simultaneously emerging: the growing ability of digitally raised youth to deal, at least under some circumstances, with moving their attention across multiple streams of information without diminished performance. There is by now a long and complicated body of research on task switching or attention switching, usually conducted with adults. Although previous studies by Poldrack and others have provided compelling evidence of the inability of most humans to switch without considerable "brain costs" (i.e., to their ability to process anything in depth), one of Poldrack's recent studies indicates that digitally raised youth can do this *if* they have been trained sufficiently in one of the tasks. If our children become far better than most adults at handling multiple sources of information, they will possess skills increasingly important for many future jobs. In other words, without necessarily preparing them to become a generation of air traffic controllers, they may well become more capable than their parents in learning to attend to and perform skillfully across attentional distractions within constraints—the details of which we need to research and understand rigorously and systematically. This is especially important, since many of them say that when they read on a screen, they are 90 percent likely to be multitasking and only 1 percent likely to multitask when reading on print media.

We are perched on a cusp between the promise and delivery of ever greater contributions by our digital culture to every aspect of our lives (including their extension) and a dawning realization of the unanticipated consequences that accompany them. Research by Steiner-Adair, Hallowell, and a growing number of others points to the need for far more in-depth research into the varied effects of the overwhelming digital claims on many of our children, particularly on cognition.

MEMORY IN A GRASSHOPPER MIND

If for me this needed research begins with attention and reading, it is because that is what I know best and where the first major cognitive impact may be most visible. It is also where we may use both science and technology to have the greatest chance for positive change. If attention in the young child, which is spasmodic and exploratory by nature, becomes all the more attenuated because of constant input, those of us who are researchers have to figure out the downstream effects on memory and other aspects of cognitive development. One question revolves around children's ability to hold things in working memory, one of the most important variables in learning literacy and numeracy. The writer Maggie Jackson has a great digital simile for thinking about working memory: "Our working memory is a bit like a digital news crawl slithering across Times Square: constantly updated, never more than a snippet, no looking back." Now consider the fact that when we adults watch television news, we are often unable to listen to the broadcaster and simultaneously read the news crawl with anything close to optimal comprehension of either content. How much more will working memory change in young children if too many stimuli are always vying for their attention? We need to know.

The second issue involves other forms of memory. If changes in working memory begin to occur, changes in long-term memory would also be predicted. If both are changed, we would predict downstream effects on children's building of their background knowledge. The latter, in turn, would impact the development and deployment of multiple deep-reading skills in the formative period of the young reading circuit.

Indirect relevant evidence is growing and comes from several sources. One of the most illustrative and earliest examples of children's online "grasshopper mind" behavior was found in a study by the Dutch researchers Maria de Jong and Adriana Bus in the early 2000s. Although e-books were far less advanced then than now, the basic choices were similar enough: the children listened either to a nonenhanced text that was read to them or to a text that was enhanced with various attention-attracting options. The four- and five-year-old Dutch children voted with their feet and hands, not their prefrontal cortex. They played with all the added elements, attended randomly to the text, and were less able to follow the narrative or remember the details than when they listened to the nonenhanced text. In other words, the number of stimuli vying for children's attention affected their memory, which affected their comprehension.

Recent studies solidify this intuitive finding. The Joan Ganz Cooney Center and the MacArthur Foundation Digital Media & Learning program have produced a series of extremely important studies and reports in the last few years about the effects of technology on children. In a study very similar in format to the Dutch research, Cooney researchers compared the effects of print books with e-books and enhanced e-books on children's literacy skills. Not unlike a growing number of other current researchers, including new work by the developmental psychologists Kathy Hirsh-Pasek and Roberta Golinkoff, they found that multiplying distractions within the enhanced e-books often as not impeded comprehension: "The highly enhanced e-book often distracted beginning readers from the story narrative. . . . In short too many bells and whistles attached to otherwise engaging technologies were not helpful to building stronger reading skills."

The inability of the younger children in these studies to reconstruct a narrative or remember its details may remind you of the older students in Anne Mangen's findings in the last letter. Recall that those older students were more likely to fail to remember the sequence and details of the plot of a passionate love story when reading on a screen rather than in printed form. Both of these findings suggest possible changes in digital readers' relationship between attention and different forms of memory, again with a potential downstream effect upon children's comprehension and their deeper thinking about what they've read. Israeli scientist Tami Katzir found just that in a large, important study of fifth-grade children. She found significant differences in reading comprehension for children reading the same story in print versus screen. Despite most children's expressed preference for digital reading, they performed better in print form in comprehending what they read.

What remains missing in all of the growing research to date is the "smoking gun" depicting the specific developmental relationships between and among continuous partial attention, working memory, and the formation and the deployment of deep-reading processes in children. Let's begin with the first three relationships. How will memory and background knowledge be affected by the digital child's expectation that there will always be effects attending to multiple pieces of constantly incoming information? And even more than for the expert reader, we must be resolute in our efforts to probe and understand the consequences of the fact that increasingly, our kids are relying more and more on external sources of knowledge such as Google and Facebook. I have several hypotheses.

When Expectation Engenders Defeat. Years ago, as a budding researcher, I presented my first formal research talk at an international neuroscience conference in Italy. Afterward, a famous British researcher, John Morton, wanted to talk to me about his supportive memory research. But first he asked me to do a little experiment: to repeat numbers back to him, in what was basically a common memory task, but he didn't tell me that. He gave no hint as to how many numbers he would be giving me. He just droned on. In reality, he gave me only seven plus or minus two numbers each time, but I didn't realize that. Rather, I expected him to give me more and more numbers each time to test the capacity of my working memory—and I froze. I could no longer repeat back even seven digits, because I expected more to come that would be impossible for me to process. I was mortified. Thirty years have passed, but the intimidating Professor Morton primed me for seeing how expectations can influence the use of our working memory capacities.

That emotion-laden (always good for long-term memory) episode leads me to hypothesize the following: there may be an incremental diminishment of the use of working memory for children due to what they perceive as the impossibility of trying to remember all the information typically presented on a screen that often moves on. Remember, the "set" we use to read on the screen medium bleeds over into our reading in print. Because children so often associate screens with TV and movies, the question emerges whether their perception of what is presented on a tablet or computer screen is being processed unconsciously like film, thus making the many details and different stimuli on the screen appear impossible to remember. So they don't. Along the same lines, older on-screen readers may be making less use of their available working memory because they, too, are processing text increasingly as if it were film, which they would never try to remember in the same way.

The Effects of Attending to Multiple Stimuli. If this speculation is true, two familiar consequences would be predicted: First, the sequence and details of the narrative would be less actively processed, thus affecting readers' memory of them. Second, the recursive dimension—the fact that when we read a physical book or paper, we can go back and read what came before—would be less invoked on-screen, where the physical space for words is as ephemeral as the continuously moving presentation of images in film. In Maggie Jackson's terms, on the screen there is "no looking back." Thus, the recursive dimension in written language would be perceived as less important than it is.

· In cognitive development terms, recursion aids looking back, which aids children's monitoring of what they comprehend, which helps them rehearse the details more in working memory, which helps them consolidate what they learn in long-term memory. If they are unconsciously processing information on the screen more like film, the plot's details would appear more evanescent and less concrete. Quite literally, the sequencing of those details would blur in memory, just as they seemed to in Mangen's older subjects and, most likely, in much younger children, too.

I take no credit for this speculation, if it proves true. The University of Chicago historian Alison Winter wrote a thought-provoking history of the role of memory in the twentieth century. She argues that our cultural inventions such as film and tape recorders and computers have changed the tasks we place on our memory and, intriguingly, serve as powerful metaphors for explaining how memory works in any historical era. She asserts that most of us still believe that the "pictures" we retrieve from our memory are what they are without reference to the nature of the cameras that took them. In my expansion here of her thinking, I am hypothesizing that film both provides a helpful metaphor for explaining what may be occurring in a child's working memory, and also may itself have become a physiological habit of mind for viewing anything upon a screen. The upshot would be less effective uses of various forms of memory in today's children, but not necessarily unalterable changes, at least at the start of childhood.

There is some support for this hypothesis in the British psychologist Susan Greenfield's work. Like Mangen, she emphasizes how the characteristics most common to narrative, such as an orderly sequence with a nonrandom cause-and-effect chain for plot events, may go by the wayside when children are processing on the screen: "While narratives are the *sine qua non* of books, they are far from guaranteed on the Internet, where parallel choice, hyper-texting, and randomized participation are more typical." Further, she asks if our screen inputs "arrive in the brain as images and pictures rather than as words, might it, by de-fault, predispose the recipient to view things more *literally* rather than in abstract terms?"

Whether or not the mismatch between screen and nar-rative will contribute to changes in both working memory and abstract thought will require much more in-depth re-search. The questions about their influence on children, however, will only become more critical for society over time, particularly when they are connected to how children use their consolidated memory to build their storehouse of background knowledge and to make critical judgments about the verity and veracity of what they see on a screen.

Our Children's Internalized Knowledge

At the heart of both deep reading and cognitive development is the profoundly human capacity that allows children to use what they already know as the basis for comparing and understanding new information to build ever more concep-tually rich background knowledge. Let me elaborate with two examples: one from your past and one from my pres-ent. Recall the Curious George story in which the beloved, mischievous monkey hitches a ride on some escaped (well, stolen) balloons and flies them through the sky. When he looks down at the ground far below, he laughs out loud at the way the houses look "like little doll houses." Children familiar with doll houses, their diminutive size and appear-ance, will begin to understand something new: that things look different, smaller, from great heights. The concept of pictorial depth perception begins with such comparisons.

But comparisons like these are useful only for children when there is a knowledge base from which to compare. Recently I visited a group of lively children in a remote part of Ethiopia where there were neither schools nor electricity nor running water nor floors of any kind. As part of our work on global literacy, I showed the children a picture of an octopus. They laughed. They had never seen or heard of such a creature, and no interpreter's attempts to explain its ocean home would ever help. Our original plan to use apps with stories about mermaids and other sea creatures flew out a non-existing window. The ocean made no sense for children who daily had to walk two hours each way to retrieve water, nor would sailing through the sky on another unknown thing, a balloon.

The act of making analogies is the great conceptual link between what is known and what is not yet known, but it is a complicated entity in children's development that is influenced by whatever the environment does or does not provide children. For many children in Western culture, that environment is providentially rich in what it gives, but paradoxically today, it may give too much and ask too little. Maggie Jackson made the thought-provoking point that when there is too much information overload, the building of background knowledge actually becomes more difficult. Like my speculations about a child's working memory, she argues that because we are given so much input, we no longer expend the necessary time to rehearse, make analogies, and store incoming information in the same way, which affects what we know and how we draw inferences.

The time needed to process what we perceive and what we read is profoundly important, whether in building memory, in storing background knowledge, or in every other deep-reading process. The literary critic Katherine Hayles sharpens this crucially important point. She emphasizes that though we have extensive evidence that digital media are increasing the volume and tempo of visual stimuli, we fail to reckon with the fact that the increase of tempo means that there is a correlative decrease in the time that the viewer has to respond. If we relate this insight to the deep-reading circuit, less time to process and perceive means less time to connect the incoming information to one's background knowledge and thus less likelihood that the rest of the deep-reading processes will be deployed.

Or Developed. As Eva Hoffman wrote about adults, our computer-based sense of time is "habituating us to ever faster and shorter units of thought and perception." In children, the convergence of more information and less time to process it may well pose the greatest threat to their development of attention and memory, with serious downstream consequences on the development and use of more sophisticated reading and thought. Everything in the deep-reading circuit is interdependent. If children are building less knowledge because they are learning to rely ever more heavily on ever more external sources of knowledge, such as Google and Facebook, there will be significant and unpredictable changes in their ability to make analogies between what they already know and what they are reading for the first time and draw accurate inferences. They will only think they know something.

This may sound familiar to you. Certainly it would have to Socrates, who worried out loud that if his students over-relied on a "papyrus that could not talk back," they would have only the illusion, not the reality, of personal knowledge. Variations on this Ur-theme have punctuated the last 150 years, as writers and filmmakers have questioned our increasing reliance on various forms of technology. Both Tom Hanks, playing an astronaut in *Apollo 13*, and Matt Damon, playing a botanist in *The Martian*, lose the ability to rely on technology and can survive only because they have the ability to rely on their own knowledge. In the first quarter of the twenty-first century, our children need to be educated like those fictional scientists from kindergarten through high school to develop both technological acumen and deep stores of internalized knowledge.

Thus, my twenty-first-century variation of Socrates' worries involves several interrelated questions: Will our culture's continuous flow of information and distraction alter or diminish young children's attention and memory? Will the fact that most "answers" are immediately available online cause older children to exert less effort to learn things for themselves? If either of these is true, will our youth develop such a passive response to knowledge that eventually the store of what they know and their ability to connect it through analogy and inference will be depleted?

If any of these scenarios becomes reality, will such changes alter other deep-reading processes, particularly empathy, perspective taking, critical analysis, and the more verbal forms of creative thought in the next generation? Can more visual forms of knowledge compensate for such losses and even provide alternate vehicles for the development of these critical skills? We tamper with our youths' intellectual development when we teach them to rely too heavily, too young, too soon on outside sources of knowledge. We also impede their progress in a digital culture when we teach them to rely too heavily, too long only on the traditional forms of what they and we already know. The intellectual development of our children depends on finding an evolving, thoughtful balance between those two principles.

The technology-averse Socrates is not my only companion in such thoughts. During an interview with Charlie Rose, Google CEO Eric Schmidt cautioned, "I worry that the level of interruption, the sort of overwhelming rapidity of information . . . is in fact altering cognition. It is affecting deeper thinking." I hope that Mr. Schmidt does not rue the day he ever said this, but I am grateful to him for his honest articulation that gets to the heart of my worries.

Will Altered Cognition Alter Deep Reading and Deeper Thinking?

Catherine Steiner-Adair titled her book *The Big Disconnect* to underscore her hope that parents can help their children disconnect from digital overuse. I am sure she would agree that an equally time-sensitive "disconnect" concerns confronting the subtle move away from children's building and relying on their own intelligence when they discover the ease of their access to external sources of knowledge. The psychologist Susan Greenfield took this position to its furthest point in a thought experiment: "Imagine in the future people become so used to external access for any form of reference that they have not internalized any facts at all, let alone put them into a context to appreciate their significance and understand them."

All of these questions and concerns may seem to contradict the intellectually visionary work on the future of intelligence as conceptualized by futurists like Ray Kurzweil. Through his work and his extraordinary inventions, Kurzweil envisions a future in which human intelligence will become continuous with artificial intelligence (the *singularity principle*), allowing us to develop exponentially expanded intellectual capacities.

Regardless of the ethical and personal-social issues involved in such visions of the future, whether the coming generations will develop highly sophisticated analogical, empathic, critically analytic, and creative capacities or not is our responsibility now. No self-respecting internal review board at any university would allow a researcher to do what our culture has already done with no adjudication or previous evidence: introduce a complete, quasi-addictive set of attention-compelling devices without knowing the possible side effects and ramifications for the subjects (our kids).

Tristan Harris is a Silicon Valley technology expert whose knowledge about the "persuasion design" principles in various apps and devices led him to become an outspoken critic of how features based on these principles are intentionally selected to addict users. Josh Elman, another Silicon Valley expert who applauds Harris's efforts, compares the use of the addictive features of various devices to the tobacco industry's use of addiction-forming nicotine before the link with cancer was discovered. The founder of the advocacy initiative Time Well Spent, Harris recently stated in interiews with PBS and *The Atlantic*, "Never before in history have the decisions of a handful of designers (mostly men, white, living in San Francisco, aged 25–35) working at three companies"—Google, Apple, and Facebook—"had so much impact on how millions of people around the world spend their attention. . . . We should feel an enormous responsibility to get this right." Most of us, including the majority of people leading and working in those three companies, would agree with this responsibility and indeed embrace it.

The responsibility starts with acknowledging that a great many of the now 1 billion cell phone users are children. These youngest among our species are intrinsically more susceptible than anyone else to persuasion principles, whether these principles tap into children's needs for social approval, or whether they involve the highly successful technique of intermittent reinforcement to influence children's increased usage. The psychologist B. F. Skinner's pigeons and our kids follow the same schedules of reinforcement to get a reward. Designers know it; casinos know it. We all should know it.

Next we need to support and conduct dispassionate longitudinal research to understand the positive and negative effects, including addictiveness, of the various media and mediums on the development of attention, memory, and oral and written language in different children. We need to connect the various, sometimes contradictory pieces of the existing knowledge about print and screen mediums and work to understand what roles each medium would play in an ideal trajectory for cognitively diverse children at different ages, in different socioeconomic environments. It is almost past high time.

You and I can hold two seemingly contradictory thoughts and not be overwhelmed by cognitive dissonance. We have reached a point where the intellectual development of our children cannot be conceptualized within a binary communication dilemma, in which one medium is intrinsically better than another. Up to this point, I have cautioned about the potential negative effects of the digital medium's affordances. Nevertheless, I am convinced that with more wisdom than we have demonstrated to date, we can combine science with technology in ways that will help discern what is best and when for each individual child from birth to adolescence, with all the mediums, devices, and digital tools at our disposal used optimally.

The stakes are far too high to cling to one side or the other. The reality is that we cannot and should not go back; nor should we move ahead thoughtlessly. Within that context, I am deeply encouraged by the work being conducted by researchers such as those in the European E-READ network, New America, the Joan Ganz Cooney Center, and the MacArthur Foundation program for their steadfast gaze on the strengths and weaknesses of our digital media and their effects on the lives of children. Like them I believe that the point of all our work is to help "build the habits of mind and the skills of critical inquiry that spur learning no matter where the text comes from, no matter whether the image is on paper or a screen."

Given the conflicting, unresolved nature of research on many of the issues surrounding the developing reading brain, I am often asked: But what should we do *now*? The next three letters are a systematic attempt to use these complex issues as the basis for imagining what I would want an ideal reading life for children from infancy to ten years to look like, given our current knowledge. And then I will leap to a future reading brain that may surprise more than a few!

Sincerely yours,
Your Author

Letter Six

FROM LAPS TO LAPTOPS IN THE FIRST FIVE YEARS
Don't Move Too Fast

Is the real barrier . . . that books can't possibly compete with exciting multimedia products for our attention? Let's face it: Screen media is the elephant in the room. True understanding of children's literacy in the twenty-first century is impossible without turning to face this creature and take a long look.

—Lisa Guernsey and Michael Levine

Books and screens are now bound up with one another whether we like it or not. Only in patiently working through this entanglement will we be able to understand how new technologies will, or will not, change how we read.

—Andrew Piper

Dear Reader,

The nursery is the "room where it happens." The first moments of my ideal reading life begin with an infant on a loved one's lap, under the "crook of an arm," where the shared touch, gaze, and experience of being read to provide the best of doors into this new, gentle realm. Before a baby can utter its first word, this never old, physical dimension of the child's earliest experience with reading connects feeling—tactile and emotional—to attention, memory, perception, and language regions in the youngest brains.

Perhaps without coincidence, early brain development gives prominence to the networks underlying feeling, even before cognition. I have always been struck by the fact that the infant's amygdala (which is involved in emotional aspects of memory) lays down its neural networks *before* networks are formed for its close neighbor, the hippocampus, the better-known store place of memory. It is a rather endearing physiological nod to Sigmund Freud, John Bowlby, Mary Ainsworth, and all those early figures in the history of psychology who emphasized the profound importance of early emotions and attachment in a child's life.

But just because infants cannot articulate their thoughts, it does not mean that they are not processing language, and from the very start. In a fascinating research study, Stanislas Dehaene and his wife, the neuropediatrician Ghislaine Dehaene-Lambertz, looked at the brain activation of two-month-old infants while they listened to their mothers speak. Using a very comfortable adaptation of fMRI, they found that the same language network that we use for listening to speech was activated in those babies. Their language network simply activated much more slowly in the earliest months of development, due to the lack of insulating myelination, which quickly enough would increase and speed transmission among neurons in the various networks. Thus, before most of us possess an inkling that babies could be listening to us, infants are making astonishing connections between listening to human voices and developing their language system.

Think how much more can happen in those regions when parents slowly, deliberately read to their children, *just to them*, with mutually focused attention. This disarmingly simple act makes huge contributions: it provides not only the most palpable associations with reading, but also a time when parent and child are together in a timeless interaction that involves shared attention; learning about words, sentences, and concepts; and even learning what a book is. One of the most salient influences on young children's attention involves the shared gaze that occurs and develops while parents read to them. With little conscious effort children learn to focus their visual attention on what their parent or caretaker is looking at without losing an ounce of their own curiosity and exploratory behaviors. As the philosopher Charles Taylor notes, "The crucial condition for human language learning is *joint* attention," which he and others who are involved in studying the ontogenesis of language consider one of the most important features of human evolution.

We can now literally see what happens to language development when a parent or caretaker reads to the child. New brain-imaging research led by the pediatric neurologist John Hutton, Scott Holland, and their colleagues at the Cincinnati Children's Hospital Medical Center provides a never-before-seen look at the extensive activation of language networks in young children who are read to, in this case, by their mothers. Hutton's group has shown how active the young brain is when it listens to stories and engages with the mother over all the things that happen to big red dogs and runaway bunnies and monkeys. Significant changes occur not only in the regions of the brain underlying the receptive aspects of language, which enhance learning the meanings of words, but also in regions underlying the expressive aspects of language learning, which enable children to articulate new words and thoughts.

The Lap Gap: The First Two Years

From both a cognitive and social-emotional perspective, I want the first two years of reading life to be the childhood equivalent of Julian of Norwich's beautiful exhortation "All shall be well, and all shall be well, and all manner of things shall be well." You see, everything counts for something when you read to your child. There is almost no end of good in what you are contributing to the various components of the reading-brain circuit. Each component part needs to be developed individually over the five years before a child learns to read. Simply consider that every single book on intrepid trains and sassy pigs, not to mention the little mouse that hides somewhere new on every page of *Goodnight Moon*, helps convey one more piece of information about the many underlying concepts surrounding those little inhabitants of childhood. All of this will lead the very young to learn how life and words work.

There is no better way for children to learn how words work. Much of my research concerns what I described in Letter Two as the "representations" of information, that are the basic elements in the components of the reading-brain circuit. When you read to your children, you are exposing them to multiple representations—of the sounds or phonemes in spoken words, of the visual forms of letters and letter patterns in written words, of the meanings of oral and written words, and so on across every circuit component. The young brain is setting down re-presentations of this information every time the child hears, sees, touches, smells books. When your toddler begs you to read *The Runaway Bunny* or *Thomas the Tank Engine* or eventually the Olivia and Madeline books over and over, she is adding one exposure after another to that information, which is exactly what strengthens and consolidates all those representations.

It is the stuff of conceptual and linguistic development (even though you might come to think it the stuff of something else entirely after the umpteenth rereading). Just remember that it is both contributing to the concepts and words your child knows already and laying the base for what comes next. Analogical thought builds within those well-worn pages, and language development flourishes. When you speak to your children, you expose them to words that are all around them. A wonderful thing. When you read to your children, you expose them to words they never hear in other places and to sentences no one around them uses. This is not simply the vocabulary of books, it is the grammar of stories and books and the rhythm and alliteration of rhymes and limericks and lyrics that are not to be found anywhere else quite so delightfully.

All of these earliest experiences provide the ideal beginnings of the reading life: first and foremost, human interaction and its associations with touch and feeling; second, the development of shared attention through shared gaze and gentle directives; and third, daily exposure to new words and new concepts as they reappear every day like magic in the same place on the same page.

YET WHY . . . .

Some of you must be asking right now, can't a child learn just as much, if not more, from the far easier repetitions of words and concepts that digital devices can effortlessly provide, not to mention the endless variety of other e-books and stories available there? Here comes the "elephant" in the nursery and the first of several concepts for you to think about in my ideas for the first steps in the reading life.

One of the characteristics that anchors the earliest experience of reading is physicality; another is recurrence: How easy is it to go back and repeat what that mischievous monkey did? And screens for young children have neither. As Andrew Piper wrote in *Book Was There: Reading in Electronic Times*, "The digital page . . . is a fake. It is not *really* there."

Physical pages are the underestimated petri dishes of early childhood. Pages give physical substance to cognitive and linguistic repetition and recurrence, which provide the multiple needed exposures to the images and concepts on those pages, which are the earliest entries in the formation of the child's background knowledge. I want children to experience the physical and temporal *thereness* of books before they encounter the always slightly removed, slightly ersatz screen. Many very young viewers all too quickly are literally and cognitively left to their own devices—to be continuously entertained by a very flat thing, which possesses neither the lap nor the voice of their most beloved persons reading and speaking just to them.

As both Andrew Piper and Naomi Baron argue, reading isn't only about our young children's brains. It involves their whole bodies; they see, smell, hear, and feel books. And with a knowing indulgent parent, they taste them, too. Not so with the lapless screen. Putting an iPad in one's mouth is just not the same. Seeing, hearing, mouthing, and touching books helps children lay down the best of multisensory and linguistic connections during the time that Piaget aptly christened the sensorimotor stage of children's cognitive development.

Second, research by developmental psychologists over the last several years shows that children who are raised with and without the so-called bells and whistles of various devices differ in early language development around two years of age. Children who receive most of their linguistic input from humans do better on language indices. Such a finding is intuitive. The input that comes from non-human sources is one step removed and not focused on one special child. Further, as engaging as such external sources can be, they rarely focus the eye's gaze or the toddler's ear on exactly what is being said or learned. In the world of the youngest children, we humans matter more. It seems almost a pity that we need to prove that.

But we do. More precisely, we need to prove what is helpful and what is not about digital media use in early childhood. For example, in a recent survey done by Common Sense Media, there is worrisome evidence that over the last ten years parents have begun to read less to their children. There are varied reasons, some old, some new. There will always be a new crop of young parents who express great surprise at what they consider the absurd notion of reading to an uncomprehending baby; they are simply not aware that the child is learning a great deal while they read. Some other parents may read less because they are deferring consciously or unconsciously to the perceived "better reader" on the screen, particularly if the parents speak a language other than English. The latter parents may never realize how important reading in their own language is to their bi- or multilingual child. And now that the tablet has become children's newest and most effective pacifier, some parents may actually read less to their children because this newest undemanding babysitter just does it for them at the end of their very busy days.

Whatever the reasons, this dip in parent-child reading was found despite all the cumulative research on its importance to later reading development. For more than four decades, one of the single most important predictors of later reading achievement has been how much parents read to their child. There are by now a spate of excellent initiatives around the world urging parents to do this, such as the US pediatricians' highly successful Reach Out and Read campaign begun by the pediatricians Barry Zuckerman and Perri Klass; the Italian Born to Read project; and Judy Koch's successful Bring Me a Book program in California and China.

The Reach Out and Read approach is backed by ample research documenting how a pediatrician's simple instructions about shared reading and the handing out of a few appropriate books at every "well visit" can change the whole pattern of parents reading to their children. Books, not apps. As Barry Zuckerman, Jenny Radesky, and their colleagues elaborate in guidelines to pediatricians and parents, physical books—not apps or e-books—are the best foundations for developing *dialogic reading*, in which parent and child form a kind of interactive communication loop that builds language and engagement. Hutton's brain-imaging data demonstrate the significant effects of this form of reading on developing language regions in early childhood.

This is all to say that before the child is two, only limited contact with digital devices is to be found in the beginnings of my ideal reading world. The devices can be present the way stuffed animals are present, neither outlawed nor ever used as a reward. Years ago, when television was the greater worry for children, my family "outlawed" television when we realized that two-year-old David was watching it excessively. He was not to blame; I was. Trying to balance my family and my professional life, I was unconsciously using the television as a pacifier substitute when I came home, much as many a parent today is doing with touch-screen devices. To correct this, from the time David was a toddler until he was ten years old, there was no more television in the home. By the time he was ten, perhaps predictably, he had become far more interested in television than any other child in the neighborhood, including his older brother, Ben, who had seen television till he was five.

I don't want to exaggerate the lessons learned here. There are many individual differences in our children, but we are all the offspring of Adam and Eve. Both young and old human beings tend to become obsessed with forbidden fruit, sometimes to the point of mystifying it and making it an object of desire. We don't need any more complexity about young children and the digital world than what is already before us.

I would like to think that there is a commonsense balance possible for pre-two toddlers that makes digital devices simply one stuffed bear among many on the shelves of childhood, but never the favorite. Before two years of age, human interaction and physical interaction with books and print are the best entry into the world of oral and written language and internalized knowledge, the building blocks of the later reading circuit.

Between Two and Five Years: When Language and Thought Take Flight Together

God made Man because He loves stories.
—Elie Wiesel

During the fleeting time from two to five years of age, children in my reading world would be surrounded by stories, little books, big books, little words, any words, letters, numbers, colors, crayons, music—lots of music!—and all manner of things that elicit their creativity, communicative abilities, and physical explorations both indoors and out.

Both musical training and various forms of physical practice such as sports and games help children learn both the discipline and the rewards of focusing their attention. Our ideal prereaders may not all become musicians or athletes, but I hope they will become little cognitive cartographers for whom each excursion into a new corner of their worlds provides fresh material for their reservoir of background knowledge and their growing experiences with words.

I would like children to have the maximum safe radius for their explorations, but for many parents this is not as simple as it sounds. Joe Frost's research shows that the radius of children's activity has shrunk by 90 percent since 1970. There are many reasons why this is so, but children build their internal background knowledge with every successful or unsuccessful exploration, as well as with every book heard, every song sung, every game played, and every rhyme and joke repeated over and over. There are many ways to increase the circumference of children's lives.

For example, just as in the first two years, I would have parents and caretakers read to their charges every day and ritualize the reading of stories every night. In this way not only do children travel in their imaginations to places very far from wherever they live, but they also become familiar with the important cognitive schemata of stories and fairy tales that will reappear over and over in their later school years. These are the stories that prepare them for their culture and teach them lifelong lessons: what it means to be a hero, a villain, or a redoubtable princess; what it means to be kind to others; how it feels when someone is unfair and unjust. The universal moral laws every culture possesses begin with stories.

Indeed, we humans are a species of storytellers. In his fascinating book *The Storytelling Animal: How Stories Make Us Human*, Jonathan Gottschall hypothesizes from a literary perspective that stories help our children and indeed all of us "practice reacting to the kinds of challenges that are, and always were, most crucial to our success as a species." Such a thought is an expansion of something that the cognitive scientist Steven Pinker also speculated upon with his argument that stories, not unlike remembered bridge or chess moves, help us face similar difficulties in life armed with possible strategies for solving them.

Just so. Much as novels provide new avenues for empathy and perspective taking in the adult reading-brain circuit, the stories of childhood provide a foundation like none other for learning the perspectives of other people and, to be sure, very lovable animals, in places that are miles or continents or centuries away. Empathy is fostered every time Martha consoles George; every time dear elephant Horton tries to help hatch what is clearly someone else's egg; every time young girls or boys or Sneetches are hurt or rejected because they are not like everyone else, no matter how hard they try. The empathy learned in stories like these expands the world of childhood and teaches an essential human value: kinship and sympathy with "other."

There is so much more going on beneath the surface here. Not unlike the research by neuroscientists that shows the excitation of both feeling and cognition when we try to understand what others are feeling and thinking, empathy is the child's platform for compassionate knowing, or what Martha Nussbaum called the "compassionate imagination." The enduring legacy of childhood's stories may begin with the simple magic woven by them, but the understanding of "others" imparted by them will stretch across the life span and, if we are all very fortunate, influence how the next generation treats its fellow inhabitants on our shared planet. Here begins the moral laboratory of human development that Frank Hakemulder describes.

Toadstools and Learning
the Secret Language of Story

Just as a moral foundation begins here, so also begins a foundation for learning words that children would never hear otherwise. Often when parents read stories to their child, they unconsciously make new words practically leap off the page. They reflexively begin to elongate certain words and animate others: "Once upon a time, there was a dark, enchanted forest where no light could enter and no creature could leave. It was in this long-accursed place that a very tiny, very shy toad lived under a very large, most unusual toadstool. The toadstool talked! Every night the toadstool whispered secrets to the toad, and every morning the toad told all the secrets to the sad princess whom he loved in vain."

No parent typically produces sentences with this many descriptive adjectives, prepositional phrases, and clauses, much less words such as *enchanted, long-accursed,* and *in vain.* This is the secret language of story found nowhere else that starts the spell with that exciting, long, tingling word *onceuponatime* and goes on to develop multiple aspects of oral and written language—like semantic knowledge (where else is a mushroom called a toadstool?), syntax, and even phonology—with no one and everyone the wiser.

A secret every child linguist knows is that no one pronounces the phonemes in words quite so distinctly as when he or she is speaking to a child. *Motherese* is a term long used by one of the liveliest and most influential child linguists of the last five decades, Jean Berko Gleason, who uses it to characterize the way we all exaggerate pronunciation, elongate words, and even use a higher pitch when we talk to a young child. "We all" includes little brothers and sisters.

I shall never forget when my five-year-old son, Ben, introduced his two-year-old brother, David, to the joys of saying "poo" and "pee" over and over in as many ways as possible. They were sitting together in a little alcove under an out-of-the-way triangular window where they thought they could not be seen. They could, however, be heard regaling each other with the unexpected ways they could string the words "poo poo" and "pee pee" together to make one nonsensical utterance after another. It was a moment of exquisite joy for them to repeat words they thought were taboo and that seemed to rhyme with just about everything they could think of. Ben and David never knew I tape-recorded the whole episode, nor did Ben realize that his excremental rhymes were giving his little brother a very fine lesson in phoneme awareness.

The research on the relationship between our children's tacit knowledge about phonemes and later reading achievement is well known; similarly between vocabulary knowledge and later reading. Less well known is older research by British experts that Mother Goose rhymes are one of the best preparations for focusing the child's attention on the phonemes of words. Whether "Little Miss Muffet" or "Hickory, Dickory, Dock" turns the child's attention to the alliterated first sounds or rhymed last sounds, what researchers call phoneme awareness is quietly developing in each child unawares—just as it developed in Ben and David's secret hideaway, the spot they called, with p-perfect alliteration, the "Poo Poo Pee Pee Place."

And just as it develops, we have discovered, in music. Research by Cathy Moritz, the neuroscientist of music Aniruddh Patel, Ola Ozernov-Palchik, and other members of our Tufts research group shows that the rhythm in music has a special relation to the development of the sounds of language, the very phonemes that are so important in later reading development.

The rhythm in music and the rhymes of language provide contributions beyond phonemes. Think what happens when you read to a three- or four-year-old child: you automatically begin to speak more clearly and more intentionally. In the process, the prosodic or melodic contour of your voice helps to convey the meanings of words to the child. You change the register of your everyday voice and become someone else. Without ever a thought, you who read to young children are effortlessly accelerating the development of many of the most important parts of the reading circuitry: the smallest sounds of words; the somewhat larger morpheme parts such as *ed* and *er*; the meanings of words; the ways that words can be used in different ways in a sentence. All of these sources of knowledge teach the child how words function in speech and story.

It is important to note, however, that only the component parts—not the connected whole of the reading circuit—are steadily developing. Unless the children are precocious outliers, such as the real Jean-Paul Sartre or the fictional Scout in *To Kill a Mockingbird*, they won't and don't need to learn to connect those parts in order to read till much later. Nor, in my ideal sequence, will they be pushed physiologically or psychologically to do so! (Do refer to what can only be called my rant on this topic in *Proust and the Squid*, if your interest is piqued.)

Protecting the Lost Time of Childhood

What should parents do about the rest of their young children's time in home and preschool environments where they are surrounded with digital devices and where their "down time" is increasingly filled with always stimulating entertainment that requires nothing of them? I wish there were a movement for the protection of lost time, when children would need very little but their imagination to make a closet door a portal and the preschool's playground the asteroid-pummeled surface of the moon. To create the space and time in childhood to do just that, exposure to digital devices will have to be introduced more gradually and more intentionally than they are now. Children should be helped to conceptualize such media as one part of their background environment, like television and music systems, but nothing to be used to consume every available empty moment of their very short time between two and five years of age.

That's easier said than done. We are all creatures of obsession, children just more so. They will become obsessed with whatever captures their attention, and there are few more effective attention getters than screens that move and buzz and bathe their senses in the hormones usually meant for fight or flight. My major fear for this developmental period is that if we do not attend—as parents and as a culture—to what makes up both the days and nights of these early years of childhood, our children and their habits will be set in screen mode.

To Connect or Not to Connect:
The Question Is What and When?

The first challenges a parent must confront concern what constitutes developmentally appropriate digital content and how long a child should be on any digital device. It is onerously difficult to figure out which apps and activities and devices are best to introduce and when for an individual child. A new parent's first exposure to the "Wild West of apps" is anything but simple. There are well over a million apps available just for the iPhone, with many thousands called "developmental" or "educational," as Lisa Guernsey and Michael Levine's comprehensive research indicates. Most of the self-labeled educational apps are not educational at all, and among those whose stated purpose is to promote preliteracy or literacy precursors for the two-to-five-year age group, precious few of them have a literacy expert involved at any stage of the design.

Summarized in their recent book, Guernsey and Levine's wise admonition is that parents always think about three C's—Child, Content, Context—before they purchase an app and consult websites specifically created to help parents evaluate the ever-growing offerings. I would add that a pain-free, delightful way for parents to begin this process is simply for them to play with their children in the first minutes after a new app is introduced. The children quickly learn to play on their own, and the parent learns just as quickly whether a particular app is engaging and worth the child's time. I am not suggesting another dimension to the "helicopter parent" phenomenon, and I am also not recommending that all the apps during the two-to-five-year period be "educational." Rather, it is important for parents to learn along with their children what engages each individual child's imagination, what develops that child's unique characteristics at different ages, and what is just drivel. Then they should simply let the child explore this medium just as he or she would a backyard or park, only not for as long!

As for just how much time and when, my hope is that parents will introduce apps and digital "toys" as something to be explored in relatively brief periods of time that increase gradually in early childhood. Discussed at more length by Catherine Steiner-Adair, a two- to three-year-old might move from a few minutes a day to a half hour, while a slightly older child gets more time, though only rarely more than two hours a day. The reality is that many children attend more formal learning environments, such as preschool, where they often have access to various digital devices during the day. The most recent statistics are that young children in this age range are already looking at screens an average of four hours or more a day.

I have no magic formula for implementing my considerably shorter, two-hour-maximum recommended allowance at home, and there will be many individual differences among children. What I seek is a day that preserves time for child-directed play and human laps and a night in which the rituals of storytelling and physical books dominate. A cumulative four or more hours of a child's day on digital devices does not easily allow for this and indeed might detract from either the child's unstructured play or the parent's shared storybook reading, particularly with real books whose pace is closer to that of the tortoise than of the hare.

There is emerging research on the latter. Increasing numbers of developmental researchers observe that when parents read stories on e-books with their children, their interactions frequently center on the more mechanical and more gamelike aspects of e-books, rather than the content and the words and ideas in the stories. Most parents are simply better at fostering language and helping to clarify concepts when they read physical books to their preschool children. As some researchers caution, the very format of the e-book may "alter the shared storybook reading, even before reading begins," with the potential for negative effects on children's comprehension and other things, too.

Adriana Bus has conducted research on shared storybook reading for many years. Her recent work demonstrates a relatively negative influence of interactive digital books on children's vocabulary and their ability to understand the content of stories. But she also includes a very recent and important caveat: when parents actively support the vocabulary of their children in interactive digital formats, there can be a more positive influence.

A promising direction for these more positive influences involves a digital genre that falls somewhere between screen and print and that is intentionally designed for human interaction between parent and child. The TinkRBook is a research tool created by my colleague Cynthia Breazeal with her PhD student Angela Chang at the Personal Robots Group at the MIT Media Lab. At the heart of this tool is a design principle called "textual tinkerability" that allows the child to . . . well, tinker with the text. For example, the child can touch a word on the screen and hear it spoken (full disclosure: this is my audiotaped voice) or see a visual image (such as a duck) and influence its actions (such as hatching out of an egg) or its attributes (such as changing the color of its feathers). In the process of interacting with the text, the child can change the whole narrative of the story. These researchers found that parents can use the interactive nature of TinkRBooks as the basis for elaborating concepts and developing vocabulary, which is the most important criticism leveled against many of the available e-books for children.

Such criticism is based in part on what does not happen enough when parents read e-books with their children and in part on what happens when e-books become a reason for parents to stop reading. For example, one of the attractive features of many current interactive storybooks is the "read to me" option. Although this feature often has very positive aspects, it appears to dissuade some parents from reading to their child at the exact developmental moment when they are needed most. Parents either feel they are less needed to read or that this option is the best babysitter in town. The worrisome consequence is that a young child can develop a far less cognitively active understanding about what reading is. When viewed by a child as yet another form of entertainment, the very attentional and reflective processes in reading we hope to promote can be lamed by passivity, a much too young example of the "Use it or lose it" principle. Such an unintended outcome would be the exact opposite of what any creative e-book or app designer has in mind or what a parent wants.

That said, it is important to note all the children who seem at home, indoors and out, with books and with tablets and who blossom with both media. For them there is less basis for the concerns raised here; they have found the desired balance. Indeed, it is the formation of the active, curious child's mind that needs to be at the center of the balance that parents of preschoolers, as well as most digital designers and researchers, seek today. We are all navigating a transition into a full-blown digital culture with many unknowns. This is the nature of transitions. It is important that we neither lurch forward with little reference to what we know nor retreat backward. With that in our minds, Cynthia Breazeal and I are now collaborating from our different perspectives on several projects such as the TinkRBook and very social robots to see if we can design digital activities that, like dialogic reading, can promote language learning and some other reading precursors, particularly for children who grow up in very different environments and will never have a book, a teacher, or a school.

Preparing All Our Children for the Future

The nursery is not the "room where it happens" for every child. There are many children who do not come from linguistically advantaged homes and for whom access to digital devices is nonexistent. Funded originally through the efforts of Nicholas Negroponte at the MIT Media Lab, Cynthia Breazeal and I helped create a global literacy initiative that eventually became Curious Learning with our colleagues Tinsley Galyean, Stephanie Gottwald, and Robin Morris. Together we are studying the efficacy of digital tablets with carefully curated/designed apps both for learning oral language and as reading precursors for children in places where there are no schools or where there is limited access to teachers, as in our South African sites, where there are between sixty and a hundred children in a single teacher's classroom. Our work began in villages in Ethiopia and has expanded to other pilot deployments in Africa, India, Australia, and Latin America. Most recently, we have begun to work closer to home with preschool age children in our own backyards in rural parts of the South.

The cautions in the last letters about learning on digital devices are informing this global and local preliteracy work and vice versa. What engages children and helps them to learn to read virtually on their own increases our understanding of literacy's early development for all children. In our future work we seek to yoke research on the cognitive impact of digital media with research by scholars such as Berkeley's Marti Hearst on the role that the human-technology interface might play in helping children learn to read, particularly children who are diverse learners or who live in adverse situations. Ongoing research by the UCLA scholars Carola and Marcelo Suárez-Orozco suggests that the growing numbers of our country's immigrant children can benefit a great deal from the way multimedia stories are able to convey important aspects about our culture, as well as teach the new language they must learn. We have only begun to connect all the related research, but our collective goal is to contribute to what the UN Sustainable Development Goals refer to as a basic human right for all the world's children: to become literate citizens whose collective potential will change the face of poverty and the reach of millions of future children.

Is There One Ideal Reading Life?

I would like to think that the principles and cautions described here for the period between infancy and age five will be useful for many of the world's children. But there are profound differences in the lives of children, some based on their particular environments and some based on their individual characteristics. Figuring out how to adapt what we know to our world's nonliterate children, for example, will be one of the great challenges in this century. Understanding how to utilize the engaging aspects of digital devices to help diverse learners is a similarly demanding, increasingly important direction in educational research. But there are also the less dramatic challenges that exist right under our noses.

I want to end this letter with a half-funny, half-sad, wholly humbling story. Not long ago, a loving, nervous, highly educated parent came to my research center to have her older child tested. As she sat in the waiting area with her second child, a very tiny, five- or six-month-old baby girl, the mother told me she had read everything I ever wrote about the importance of reading to one's child. I looked with no small interest at the very large bag of books on the floor beside her. With a very slight glance at me, the mother promptly plopped her baby onto her lap and began reading. With a voice pitched at high C and racing at perilous speeds through one of the bag's many Dr. Seuss books, she appeared intent on finishing all thirty pages of the book—an intention clear to all, including her baby. Within two minutes, the little one squirmed; within three minutes, she began to whimper, throwing her hands out in futile protest. Within four minutes, she deteriorated before our eyes. Nothing would deter this well-meaning, millennial mother from the duty she felt to read to her child as often and as much as possible. I had created a Reading Raptor.

As gently as I could, I discussed how we don't have to read a whole book or whole story every time we read to a child; that it is good to read only as much and only as fast as each child can handle; and that simple picture books supplemented by a few words from her can be as helpful to her infant as a Dr. Seuss book would be to a slightly older child.

What I wish I had said is what I now say to you, too: trust your inner mother or father or grandmother or grandfather. What and how would they read to this very new, little person? Shared attention, as Charles Taylor wrote, is the beginning of the great dance of language that joins one generation to the next, not forced attention. Knowing research about the development of literacy is a very good thing; knowing what to attend to in one's own child overrides everything I can ever say—or write—about any medium or any approach.

There are so many things we all have to learn. That is especially true for children who are about to move through the kindergarten door.

Warning: it will not be what you expect.

Faithfully yours,
Your Author

THE SCIENCE AND POETRY IN LEARNING (AND TEACHING) TO READ

There is nothing that a little bit of science cannot help. Parents and educators must have a better understanding of what reading changes in a child's brain. . . . I am convinced that increased knowledge of these circuits will greatly simplify the teacher's task.

—Stanislas Dehaene

And what do we learn from Seuss? The joy of words and pictures at play, of course, but also the best and most humane values any of us wish to possess: pluck, determination, tolerance, reverence for the earth, suspicion of the martial spirit, the fundamental value of the imagination.

This is why early reading matters.

—Michael Dirda

Dear Reader,

Between the time they are five and ten years old, children around the world begin to learn to read and enter the most exciting learning adventure of their young lives. In William James's apt description, "children who learn to read . . . take flight into whole new worlds as effortlessly as young birds," their first stop on the way to Dinotopia and Narnia and Hogwarts. Along the way they will fight all manner of monsters from dragons to bullies; they will discover all kinds of "others"; they will swoon over heroes or swear they will never swoon. But most of all they will leave their desk or their chair or their bed to discover who they might become. As Billy Collins wrote in his wonderful poem "On Turning Ten," at four he was an Arabian wizard, at seven he was a courageous soldier, and at nine he became a prince.

For all too many children, however, none of this is true. For them, the walk through the kindergarten door is the beginning of a recurrent nightmare that is invisible to almost everyone else. Depending on which scenario they experience, either children will have their own shot at the elusive American dream or they will not, with far-reaching consequences for everyone in society.

Every national and international index of how well US children are doing in reading indicates that, despite all the nation's wealth, they are failing in droves and performing considerably behind children in both Western and Eastern countries. We cannot ignore what this portends for our children or for our country. There are facts to know, whether we have children ourselves or not, and, most important, things that all of us can do about them to reclaim the potential of our country's children.

Specifically, the recent national report card (National Assessment of Educational Progress) documents that a full two-thirds of US children in the fourth grade do not read at a "proficient" level, that is, fluently and with adequate comprehension. Put in more sobering terms, only one-third of twenty-first-century American children now read with sufficient understanding and speed at the exact age when their future learning depends on it. The fourth grade represents a

Maginot Line between learning to read and learning to use reading to think and learn.

More disturbing altogether, close to half of our children who are African-American or Latino do not read in grade four at even a "basic" reading level, much less a proficient one. This means they do not decode well enough to understand what they are reading, which will impact almost everything they are supposed to learn from then on, including math and other subjects. I refer to this period as the "vanishing hole in American education" because if children do not learn to read fluently before this time is over, for all educational purposes, they disappear. Indeed, along the way many of these children become dropouts with little hope of reaching anyone's dream when they grow up.

The Bureaus of Prisons in states across America know this well; many of them project the number of prison beds they will need in the future based on third- or fourth-grade reading statistics. As the former CEO and philanthropist Cinthia Coletti has written, the relationship between grade-four reading levels and dropping out of school is a bitter, overwhelmingly significant finding. She contends that if this many children are seriously underperforming in the schools, our country cannot maintain its leading economic position in the world. Buttressing Coletti's conclusions, the Council on Foreign Relations issued a report in which it stated with no ambiguity, "Large, undereducated swaths of the population damage the ability of the United States to physically defend itself, protect its secure information, conduct diplomacy, and grow its economy."

Only a proficient reading level will ensure that an individual can go on to develop and apply the sophisticated reading skills that will maintain the intellectual, social, physical, and economic health of our country. Two-thirds or more of future US citizens are not even close.

Where Do We Start?

For these children, the first five years before they go to any school have no resemblance to the ideal life I described in the last letter. I am weary of citing the old and new studies that document the 30 million–plus instances of words that children in underprivileged families do not hear in their environments, and the numbers of books and letters they fail to see, much less hear read to them, before they are four and five years old. Money literally talks in the early language and cognitive development of our children, as demonstrated in the extensive analyses by the University of Chicago economist James Heckman and his colleagues. Simply put, the amount of money we invest in the first years of a child's life produces greater returns for each dollar spent than at any other time in the life span. The implications of all the various types of research on the developing child could not be better understood: society needs to invest in more comprehensive early-childhood programs with more highly trained professionals before the first large gaps in language and learning become permanently cemented in the lives of millions of children.

A caveat: Nonie Lesaux, a language scholar at the Harvard Graduate School of Education, rejects the term *gap* because it suggests that all we have to do is fill it and our work will be done. She is correct. Most children who are underserved in the first five years of life underperform in the next five and the next, and they continue to be underserved for the rest of their years. Unless we change the whole equation: we need to reconceptualize the time from zero to five years, the first two thousand days of life, when the component parts of the reading circuit are laid down as discussed. We need to rethink the time from kindergarten to fifth grade, the second two thousand days. This period, the focus of this letter, is when children learn to read and think in ways that lay the foundation for the rest of their lives. During this time, the baton passes formally to the schools, where three investments are needed to ensure that all our children reach their potential as contributing members of our society: comprehensive ongoing assessment from the outset; excellent, well-informed teaching methods; and coordinated emphases by all teachers on developing reading and language skills across the grades. Each requires different forms of investment.

INVESTMENT IN EARLY, ONGOING ASSESSMENT OF STUDENTS

When children step through the kindergarten door, they come in all sizes, abilities, languages, dialects, and cultures. The school's first job is to figure out who is ready to learn, who is not, and what to do about it. From the very first day the schools must be able to assess what is needed for those children who did not receive a quality preschool experience and may well be behind in language development and other precursors of reading. From the second day teachers need to know whether children who have had a high-quality preschool experience have different strengths and weaknesses that will require specific emphases before they are more formally taught to read. Everyone involved in what happens next needs to be aware of some important new research as well as some well-established older research, neither of which is sufficiently known or implemented in many schools.

An exciting new study could change business as usual in the first two days of school. My present and former PhD students Ola Ozernov-Palchik and Elizabeth Norton, along with John Gabrieli and his colleagues at the McGovern Institute for Brain Research at MIT and Nadine Gaab at Boston Children's Hospital, just finished one of the largest reading prediction studies ever conducted. These are the kind of studies that help us predict who will go on to do well in important subjects such as reading and math and why, and who will need to be carefully followed.

Our group studied well over a thousand kindergarten children from every economic circumstance and from all over New England. Each child was tested on a large battery of educational measures. The results highlighted two facts, one unsurprising and one potentially transformative. First, American children bring with them profound cognitive and linguistic differences the first day of formal schooling; not a surprise. Second, these differences fall into fairly discrete groupings that predict how the children will achieve in reading later in school. This could change the trajectories of many children.

Specifically, six developmental profiles emerged that can help teachers and parents understand what each group needs and how each group learns to read best from the very start. Two of the profiles comprise children who are either average or very much above average and will need only good instruction to excel. Another group has difficulty with letters and sounds and may well come from environments where there is little exposure to the alphabet or the English language. We can redress these issues fairly straightforwardly. Some children in this group, however, may have more rare visual-based difficulties that need further testing.

Three of the profiles comprise children who we know will go on to be diagnosed with some form of reading disability or dyslexia. The brain organization that gives children with dyslexia significant advantages later in their lives—in areas such as art and architecture, pattern recognition in radiology and finance, and entrepreneurship—disadvantages them during their first years of learning. There are few discoveries more important to those of us who study dyslexia than to be able to predict it before the child has to endure ignominious, daily public failures before peers, parents, and teachers. Indeed, there is little more destructive for a six-year-old child than to suddenly think that he or she is dumb because everyone else can read but him or her, whether the reason is biological or environmental or in some cases both.

By assessing struggling young readers early on, we can prevent some of the emotional detritus that often characterizes their reading experiences. In the process we can save society large sums of money by preventing the need for some prison beds and by preserving the spirit of children with dyslexia, who can then go on to become some of our most creative members and successful entrepeneurs.

The critical point here is that we are now on the threshold of being able to predict highly specific reading trajectories of young children before they ever begin to read. Other researchers at UCSF School of Medicine, led by Fumiko Hoeft and Maria Luisa Gorno-Tempini, are working to refine our batteries and profiles, but already such information in the hands of trained teachers could prevent some reading problems, ameliorate others, and deliver intensive early intervention for the children most at risk for dyslexia. Nothing in reading acquisition is more important than beginning systematic, targeted intervention as early as possible.

This research helps all children, not just those with more obvious learning challenges. The prediction battery also demonstrated the tremendous developmental variability at this age among the largest group of the more typically developing children. Some children, particularly boys, show no obvious areas of weakness in their profile but are simply not yet ready. Understanding this group requires more in-depth evaluation (to ensure that there are no underlying weaknesses) and also more reasonable expectations for our children than is sometimes the case. Too many schools have school administrators who are under such pressure for their children to do well in later grades on the publicly recorded state tests, that they pressure their teachers to push reading acquisition earlier and earlier in the kindergarten curriculum. The Johns Hopkins pediatric neurologist Martha Denckla vehemently argues that we may be causing as many impediments to reading as preventing them by our push to get every child reading before they leave kindergarten.

The British reading researcher Usha Goswami reinforced this conclusion in a study of reading practices in Europe to establish when reading instruction ideally should begin. She found that in the countries that introduced reading later, reading developed with fewer problems for the children. In other words, European children who began instruction in what we would consider first grade acquired reading more easily than those who began a year earlier.

These results are, to be sure, confounded, because there is more orthographic regularity in the languages of the countries that introduce reading a year later than we do. Nevertheless, there are sound physiological and behavioral reasons why some children are simply developmentally not ready in kindergarten. The bottom line is that fears about third-grade state test results in the United States should never dictate decisions about when whole kindergarten classes receive instruction for reading. Some children are pushed to read too hard too soon, before they are developmentally ready. Some children read well before they end kindergarten or even enter it. Others are sent to first grade to receive the intervention du jour in their school that is inappropriate for their specific learning profiles. Perceptive, well-trained teachers, excellent prediction tools, and better-targeted, evidence-grounded interventions are our best defense against any of these all-too-common errors that derail children's development.

INVESTMENT IN OUR TEACHERS

Over the last half century our society has gradually handed over to teachers, arguably its most idealistic members, all the ills that society itself could not "fix," particularly the pernicious effects of poverty and stressful environments on early child development. Every school community should watch the documentary *The Raising of America*, by the filmmaker Christine Herbes-Sommers, for an honest, astringent accounting of how these effects last a lifetime. Most teachers, however, receive neither sufficient preparation in their graduate schools, nor the professional development afterward to meet the escalating range of challenges that confront them in today's classrooms—from an increasing range of attention and learning challenges, to the particular needs of increasing numbers of dual- and multilanguage learners, to the uses of technology in the classroom.

Knowing how to introduce all children with their many differences to the reading life today is as complex a set of knowledge bases as any engineer, rocket scientist, or saint is ever called upon to use. Today's teachers need to be prepared with new knowledge, particularly about the reading brain and its implications for how we teach teachers and children. As Stanislas Dehaene emphasized, what we know about the reading-brain circuit can enrich the development of teachers' understanding, especially concerning the merits of different forms of reading instruction. It may ultimately bridge one of the most intransigent debates about methods of teaching, the so-called Reading Wars.

The Debate That Should Never Have Been. By and large, twentieth-century educators were trained within two strikingly different approaches to the teaching of reading. In the approach called *phonics*, reading instruction begins with children understanding the basic elements underlying the alphabetic principle: that words consist of sounds or phonemes and that these sounds correspond to letters in the alphabet, with rules that are to be learned as the entry to reading. Teaching is explicit, and the emphasis moves from a foundation in English phonemes and letters to systematic rules about connecting letters to sounds and about decoding different types of words.

In the approach called *whole language*, learning is to be implicit: the rules of decoding are to be inferred or figured out by the child, with little to no explicit instruction in decoding or emphasis on the phonemes in English. The emphasis centers on engagement in stories, authentic literature, word meanings, and the imagination of the child to the exclusion of phonic principles. Indeed, some earlier professors of education erroneously labeled phonics methods "kill and drill" and characterized teachers of phonics approaches as less progressive and child-centered.

Both approaches came to be favored by very fine teachers, many of whom maintain to this day a loyal, sometimes even zealous, belief in the methods they were originally trained in during their teacher certification programs. Why each of these approaches ever came to exclude the emphases of the other is one of the great unfortunate errors of the twentieth century. Unfortunately it continues. Even though there is a movement toward "balanced reading," the too-frequent reality is a thinly veiled variation of the whole-language approach with a cursory, unsystematic nod to phonics principles. This is all too understandable but lamentable.

Extensive, federally funded research studies unambiguously support the importance of children's learning to read through the explicit teaching of the basic principles of decoding. While clearly supporting phonics principles, these results never mean neglecting engagement with literature, as the more recent, hopeful emphasis on what are called *common core principles* for the education of our children indicates. Although difficult to implement, the revised common core standards embody the importance of science and imagination for teachers and students throughout the school years.

The problem is that neither evidence from science nor the experience of being unable to teach many children to reach functional literacy levels has proven sufficient to many teachers across the United States and Australia, still bound by their allegiance to whole-language methods. In one of the best new overviews of research on reading and this topic, Mark Seidenberg memorably described these methods as "theoretical zombies that cannot be stopped by conventional weapons such as empirical disconfirmation, leaving them free to roam the educational landscape." Such a situation is a double waste: it wastes the unquestionably idealistic intentions of the whole-language teacher, and it thwarts learning to read for many children, especially those with reading or learning differences or dual language needs. That said, neither Seidenberg nor I would ever take one minute away from the whole-language teacher's time spent on bringing words, stories, and a life of reading enjoyment to children, as long as they do not preclude a systematic, informed approach to learning the phonemes of the language, the alphabetic principle, and decoding rules.

From a cognitive neuroscience perspective the repetition

fostered in the latter approach provides children with the multiple exposures they need to learn and consolidate the rules for letters and their corresponding sounds and increase their knowledge of words, stories, and literature. Repetition fosters the growth of high-quality representations from phonemes and graphemes (letters) to word meanings and grammatical forms. As a very old teacher once said, "Most times, the lower rungs of a ladder are the best ones for learning to climb. I always hate to ask a child to jump to the top rung without them." All of the rungs are important if we are to prepare children to become fluent readers who use both their imagination and their analytical capacities.

Furthermore, knowledge about the reading brain can help teachers of every method see what rungs in the ladder may be missing in how they teach children. The reading circuit activates everything it knows. *So should our teaching during the entire five- to ten-year age period.* Within such a perspective, teachers of children from five to ten years of age would give ample, explicit attention to every component of the reading circuit: from phonemes and their connections to letters; to the meanings and functions of words and morphemes (e.g., the smallest units of meaning) in sentences; to an immersion in stories that require ever more sophisticated deep-reading processes; to the daily elicitation of the children's own thoughts and imagination in speaking and writing.

In such a way nothing relating to cognition, perception, language, affect, and the motor regions is neglected. At no time during the primary grades should any of these components be neglected and not given their full place in instruction. Learning the meanings and grammatical uses of words in increasingly complex sentences is important in first and third grades. Learning about new letter patterns that always reappear and help us figure out the meanings of words is important in both first grade and fourth grade. Over time—by the third and fourth grades—these lower-level, basic circuit components need to be so practiced and automatic that children can turn their attention to ever more sophisticated comprehension processes, beginning with expanding their background knowledge and ending with the elicitation of their insights and reflections.

This is the basis of fluency and also the best way of acquiring it. Fluency is not simply about the speed of decoding, an assumption that has led to the common but insufficient practice of having children reread a passage over and over again. Think back to the Cirque du Soleil image: Each ring has to be fast enough in and of itself so that it can pass its information forward to the other rings. Only when each of the rings is fast enough to work in tandem with the other rings can time be allocated to comprehending what is read and having feelings about it, too.

We now have extensive evidence that an approach to reading that emphasizes all of these parts of the reading circuit benefits many children. A decade of research by Robin Morris, Maureen Lovett, and my research group has been funded by the National Institute of Child Health and Human Development. This body of randomized treatment-control studies (the gold standard of research in medicine and education) demonstrates that when the major components in the reading circuit are explicitly emphasized—the earlier the better—children become more proficient readers, even when they begin with significant challenges like dyslexia.

Further, as new work in English by Melissa Orkin, in Hebrew by the Israeli scholar Tami Katzir, and in Italian by Daniela Traficante illumines, fluent reading involves knowing not only how words work but also how they make us feel. Empathy and perspective taking are part of the complex woof of feelings and thoughts, whose convergence propels greater understanding. All young readers need to be able to look at a question such as "Did Horton lay the egg he sits on?" and smile with recognition and affection.

Emphases on the multiple aspects of words are not only critical for fluent, proficient reading in the two-thirds of our children who are failing at this moment, they are the bridge that connects the decoding of words to deep-reading processes. Rereading the same stories and sentences over and over again is helpful practice for gaining speed on a particular text, but it will never prepare children to connect concepts, feelings, and personal reflections. Deep reading is always about *connection*: connecting what we know to what we read, what we read to what we feel, what we feel to what we think, and how we think to how we live out our lives in a connected world.

The importance of forming these connections was brought home to me years ago by Martha Nussbaum's *Cultivating Humanity*: "Education for world citizenship needs to begin early. As soon as children engage in storytelling, they can tell stories about other lands and other peoples . . . [they can] learn . . . that religions other than Judaism and Christianity exist, that people have many traditions and ways of thinking. . . . As children explore stories, rhymes, and songs—especially in the company of the adults they love—they are led to notice the suffering of other living creatures with a new keenness."

Stories are one of humanity's most powerful vehicles for making lasting connections to people we will never meet. To feel like Charlotte about Wilbur's plight in *Charlotte's Web*, to identify with Martin Luther King, Jr., in *Martin's Big Words: The Life of Dr. Martin Luther King, Jr.*, or with Ruby Bridges in *Through My Eyes* prepares children to empathize both with their neighbors and with people

whose lives take place around the world or across the proverbial railroad tracks. Recall James Carroll's transformation after he read Anne Frank's *Diary*. Think about the
ways you have been changed by fictional characters such as
Celie in *The Color Purple* and Hamlet and the real lives of
people such as Eleanor Roosevelt in her autobiography and
James Baldwin in *I Am Not Your Negro*. Whatever our
age, we can be changed by the lives of others if we learn
to connect the whole of the reading circuit with our moral
imagination.

INVESTMENT IN THE TEACHING OF READING ACROSS THE SCHOOL YEARS

None of this ends in the primary grades. If we are to change
the desultory results of the NAEP's report card for our nation and, more important, change the lives of the droves
of children lost from fourth grade on, teachers in higher
grades need to receive training in teaching children who
do not read at grade level. I referred earlier to the double-
edged Maginot Line in fourth grade. It is the moment when
reading changes and when the content of what is to be read
becomes ever more demanding in complexity. It is also the
moment when higher-grade teachers assume that children
have already been taught to read and should no longer need
assistance. It is a false, destructive assumption that must be
changed, beginning with reconceptualizing teaching certification programs.

The education of my son Ben illustrates this point all
too poignantly. Ben was and is dyslexic in the most proto-
typical ways: creative, wonderfully intelligent, and sensitive to the daily insults one must deflect when one is unable
to do what every other kid can: read. The fourth grade was
his worst of times, despite the fact he and his brother, Da-

vid, attended a very nurturing Friends School, which laid a foundation for equality and fairness like few schools I have ever seen. Ben was one of five boys who were not reading at the same level as the rest of the class. They were a pack of trouble, or so thought their very well-meaning, ardently feminist teacher whose enthusiasm for girls' education and disapproval of boys' shenanigans went too far, or so thought my son and his friends.

With all the sense of rightness for just causes that the school emboldened, Ben and his buddies organized a petition against the "sexist unfairness" of the fourth-grade teacher toward boys and the unjust treatment they received over their homework. After they delivered their petition to the school principal—with quite a few signatures, it should be noted—they returned to their classroom with that momentary sense of vindication that the righteous surely deserve—until it is replaced by the reality of an irate fourth-grade teacher.

She felt blindsided. What she was blind to was the fact that each of the boys was acting out because he was unable to meet her expectations, worthy though they were, of the fluent reading comprehension that fourth-graders were supposed to have attained. She never felt that there was a need to teach them more reading skills, for that was the jurisdiction of the primary grades. She had little patience for things outside what she had been taught. I wish I could say there was a happy ending. There was only a decision by the parents of four out of the five boys, including Ben, that the children would need schools better equipped to deal with the variety of their diverse learning challenges.

What that fourth-grade teacher lacked was not compassion. What was lacking was the kind of knowledge that would have given her the basis for understanding that not all kids come to or leave grade four able to read fluently;

the kind of training that would have enabled her to teach older children to do just that; the motivation to work until no child in her classroom failed. The teaching of reading is hard, full of pitfalls, with obstacles all along the way until children reach the level of proficiency that allows them, whatever their learning trajectory, to pass over from the text to their own thoughts and return enriched. In my ideal reading world, it happens by third or fourth grade. In the real reading world of schools in the United States, it does not.

But it can. There are no simple solutions, particularly given the increasingly complicated needs in today's class-rooms. Along with greater knowledge, better training, and full buy-in by our primary and elementary school teachers and administrators, we can come closer to an ideal reading life for many more nonproficient children. But we have to think outside the box. A large, ongoing initiative by the Strategic Education Research Partnership (SERP) is a case in point. Led by a former editor of *Science*, Bruce Alberts, philanthropists such as Cinthia Coletti, and scholars such as Catherine Snow, this multidisciplinary initiative helps teachers from different areas across the middle grades around the country. One prominent aspect of their work helps schools prepare their students with a shared corpus of words and concepts that will advance literacy and critical thinking across disciplines. These words are reinforced and elaborated by every teacher in every grade: for example, through stories in language arts classes, through historical facts in social studies classes, and with new meanings in math or science classes. By the time they graduate, students will have acquired a repertoire of core concepts and words that will serve as a foundation for the rest of their learning.

We need to invest in providing teachers across the en-tire elementary school grades with new knowledge—from research on the reading brain's implications for early assessment, prediction, and more individualized multidimensional methods of teaching reading, to schoolwide initiatives on reading and language, to digital-based learning tools. Our twenty-first-century children must develop habits of mind that can be used across various mediums and media. Thus, our teachers also need far more knowledge than most now possess about the ways in which digital learning can contribute to solving the present crisis of our students—without exacerbating the increasing problems of attention, background knowledge, and memory. That requires its own letter, and it may well surprise those of you who by now privately regard me as a closet Luddite. But buckle up. We're all in for a wild ride.

Sincerely yours,
Your Author

BUILDING A BILITERATE BRAIN

Defining potential long-term problems is a great public service. Overdefining solutions early is not.
—Stewart Brand

The depth of the challenge: there are no well-known, tried-and-true curricula for teaching young children how to use online information or think critically about visual information while they are simultaneously learning to read print, listen and speak in full sentences. This is uncharted terrain.
—Lisa Guernsey and Michael Levine

Dear Reader,

I have little doubt that the next generation will go beyond us in ways we cannot imagine at this moment. As Alec Ross, the author of *The Industries of the Future*, wrote, 65 percent of the jobs our present preschoolers will hold in the future haven't even been invented yet. Their lives will be extended much beyond ours. They may well think very different thoughts. They will need the most sophisticated armamentarium of abilities that humans have ever acquired to date: vastly elaborated deep-reading processes that are shared with and expanded through coding, designing, and programming skills, all of which will be transformed by a future that none of us—from Stewart Brand, Sundar Pichai, Susan Wojcicki, Juan Enriquez, and Steve Gullans to Craig Venter and Jeff Bezos—can now predict.

Building the kind of pluripotential brain circuitry that can prepare the youngest members of our species to think with the knowledge and cognitive flexibility they will need is one task that we, their guardians, can attend to in the short time we share the planet. Whatever their next iterations, the future of the reading circuit will require an understanding of the limits and possibilities of both the literacy-based circuit and digital-based ones. This knowledge involves examining the often contradicting strengths, weaknesses, and sometimes opposing values that characterize the processes emphasized by different mediums and media. We need to study the cognitive, social-emotional, and moral impact of the affordances of present mediums and work toward the best possible integration of their characteristics for future circuits. If we are successful, we will recapitulate in our next generation's physiology Shakespeare's great lesson about love: "Mine own, and not mine own."

The philosopher Nicholas of Cusa can help us. He believed that the best way to choose between two seemingly equal but contradicting perspectives—what he called the "coincidence of opposites"—was to assume the stance of *learned ignorance*, in which one strives to thoroughly understand both positions and then goes outside them to evaluate and decide the course to be taken. Knowledge about the reading brain and the directions of its future iterations requires yoking research from multiple disciplines—from cognitive neuroscience and technology to the humanities and social sciences. No one of these disciplines is sufficient to make the kind of decisions we need to make; each of them adds something essential to the combinatoria of knowledge we need to develop Nicholas of Cusa's stance of learned ignorance. Within this context I propose the development of a biliterate reading brain.

A Developmental Proposal

We start by building a childhood that is not split between two mediums of communication but rather, in Walter Ong's words, is "steeped in" the best of both, with more options still to come. You already know what I think about the print medium's role and the gradual introduction to a second, digital medium in the first five years. The second five are our real challenge.

I propose a relatively simple, perhaps novel design for introducing different forms of print-based and digital-based reading and learning during the five- to ten-year age period. Its overall blueprint is based on what we know about nurturing dual-language learners whose father and mother each speak a separate language and the parent who spends the most time with the child speaks the language that is less spoken outside the home. In this way, young bilingual children learn to speak both languages well. They gradually get beyond the inevitable errors that arise in going from one language to another and ultimately are able to tap into their deepest thoughts in either language. Very importantly, during this process they learn to become expert code switchers. By the time they reach adulthood, their brains are masterpieces of cognitive and linguistic flexibility, which we can see in fascinating ways.

Many years ago, aided by insights by my Swiss friends Thomas and Heidi Bally, I created a naming speed task called the Rapid Alternating Stimulus (RAS) test, now used by neuropsychologists and educators to predict and diagnose dyslexia. Basically it asks a person to name a series of fifty well-known items in different categories, specifically letters, numbers, and colors. The person has to switch from one category to the next as fast as possible, which requires both considerable automatized knowledge and a great deal of flexibility. An unexpected finding in the various comparison studies was that bilingual adults were *faster* on these tasks than were their monolingual peers. Dual-language learners had acquired far more verbal flexibility than single-language learners had.

As shown by groundbreaking work by Claude Goldenberg and Elliott Friedlander at Stanford University and at Save the Children, bilingual and multilingual speakers have spent years going back and forth between languages. Not only are they more flexible in retrieving words and concepts, but there is some research that indicates that they are also more capable of leaving their particular viewpoints and taking on the perspective of others.

That is what I want our young nascent readers to become: expert, flexible code switchers—between print and digital mediums now and later between and among the multiple future communication mediums. My thoughts about how this would work over time are inspired by the Russian psychologist Lev Vygotsky's depiction of the development of thought and language in the young child: first separated and then increasingly connected. Thus I conceptualize the initial development of learning to think in each medium as largely separated into distinct domains in the first school years, until a point in time when the particular characteristics of the two mediums are each well developed and internalized.

This is an essential point. I want the child to have parallel levels of fluency, if you will, in each medium, just as if he or she were similarly fluent in speaking Spanish and English. In this way the uniqueness of the cognitive processes honed by each medium would be there from the start. My unproven hypothesis is that such a codevelopment might prevent the atrophy seen in adults when screen-reading processes bleed over into print reading and eclipse the slower print-reading processes. Rather, children would learn from the outset that each medium, like each language, has its own rules and useful characteristics, which include its own best purposes, pace, and rhythms.

THE ROLE OF PRINT

In the first school years, physical books and print would be used as the principal medium for learning to read and would dominate story time. That was the lesson in Letter Six: reading in print by parent and child reinforces core temporal and spatial dimensions in reading, adds important tactile associations in the young reading circuit, and provides the best possible social and emotional interaction. Whenever possible, a teacher or parent would ask questions that lead children to connect their own background knowledge to what they read; that elicit their empathy for another's perspective; that prompt them to make inferences and begin to express their own analyses, reflections, and insights.

Learning the importance of allocating time to their nascent reflective processes is anything but simple for children raised in a culture filled with distractions. As Howard Gardner and Margaret Weigel noted, "guiding this peripatetic mind may be the primary challenge of educators in the digital era." The explicit encouragement of the earliest deep-reading skills in young readers would be an antidote to the continuous temptations of digital culture: to skim quickly and move on to the next interesting thing; to be passive and conceptualize reading as one more game that entertains and is over; to skip figuring out their own thoughts. As one student opined, "Books slow me down and make me think, and the Internet speeds me up." Each would have its place; moreover, children would learn what is best for different learning tasks.

For example, during the initial introduction to print reading, we want children to learn that reading takes time and gives back thoughts that continue long after a story is finished. Just as children's natural tendency to dart from one thought to the next may be exaggerated by frequent digital viewing, the experience of deep reading can help give them an alternative mode for their thoughts. Our challenge as a society is to give digital children both these kinds of experiences. They will need concerted efforts by their teachers and parents to be sure they read fast enough to allocate attention to deep-reading skills and slowly enough to form and deploy them.

Through this five-to-ten-year period, the goal is to instill in children the expectation that if they take their time, they will have their own ideas. All children—particularly children who feel insecure due to having to learn to read— gain something in the process of this kind of thinking that sets the stage for the rest of their lives. They learn to expect something important of themselves when they think reflectively about what they have read.

Another stratagem for helping children think while they learn to read may surprise you. Learning to write by hand encourages them to explore their own thoughts at closer to a snail's pace than a hare's, particularly if their spelling still veers into the "gnys at wrk" variety. There is a growing body of research on handwriting that demonstrates that when children learn to write their thoughts by hand in the early grades, they become better writers and thinkers. From a cognitive neuroscience perspective, the beneficial cortical connections between language and motor networks are something Chinese scribes and teachers have known for centuries.

TEACHING DIGITAL WISDOM

At the same time that children are learning to think and to read in the slower medium of print, they will also be learning to think in a different way on fast-moving screens. Digital devices would be introduced as a medium for coding and programming and what the Tufts technology researcher Marina Bers calls the "playground" for learning a stunning variety of creative digital-based skills: from making graphic art and programming Lego robots to creating GarageBand compositions. Any one medium would not be given a preferential place in the classroom. In the process of learning to code, young children develop the deductive, inductive, and analogical skills that are used in all STEM (science, technology, engineering, and math) learning and that simultaneously make up the core "scientific method" processes in the reading circuit. They begin to understand, for example, that sequence matters, the main weakness exposed in digital reading by Anne Mangen's research. The importance of sequencing and other STEM processes is emphasized in the introduction to Scratch, the coding program for young children designed by the director of the Lifelong Kindergarten group at the MIT Media Lab, technology expert Mitchel Resnick, and Marina Bers. In one of the single best descriptions of coding, they wrote:

> Every child should be given the opportunity to learn how to code. Coding is often seen as difficult or exclusive, but we see it as a new type of literacy—a skill that should be accessible for everyone. Coding helps

learners to organize their thinking and express their ideas, just as writing does.

As young children code . . . they learn how to create and express themselves with the computer, rather than just interact with software created by others. Children learn to think sequentially, explore cause and effect, and develop design and problem-solving skills. At the same time . . . they aren't just learning to code, they are coding to learn.

In another part of the MIT Media Lab, Cynthia Breazeal helps children acquire a variety of coding skills through personal interactions with what have to be the cuddliest robots any child has ever encountered. She and her team demonstrate how a combination of social interactions and programming skills help children learn how to construct, deconstruct, and program robots so that they can move, whirl, and beep. In the process children understand why and how things in the digital world work. Such active forms of digital-based knowledge give children insights that cross every domain of learning. Specifically, the parallel, mutually supportive processes that children learn while coding and creating complement the processes used in learning to read in a print medium.

At some still unpredictable instant comes the moment of transition, when the child has learned a great deal within both mediums and across multiple media and is ready and probably anxious to be reading more schoolwork on screens. Exactly when this happens will depend on the child's individual characteristics, reading abilities, and environment. Understanding individual differences matters. For some children, online reading instruction can't begin too soon; for others, it needs to be phased in much more slowly over time. Mirit Barzillai, along with Jenny Thomson and Anne Mangen from the European E-READ network, is trying to confront the cognitive challenges inherent in children's reading on screens in a digital world. Both they and I are convinced that the next generation's deep-reading processes will be most endangered by the digital medium if we do not teach the proper uses of digital learning and screen reading from relatively early on, rather than leaving children to develop willy-nilly digital habits of mind that are counterproductive.

To prevent the latter, "counterskills" would be taught as soon as a child begins reading on screens. Emphases would be placed on the importance of reading for meaning, not for speed; on avoiding the well-known skimming, word-spotting, zigzag style of many adult readers; on regular monitoring of their comprehension while reading (checking for the plot's sequence and "clues" and rehearsing their memory for details); and on learning strategies to ensure that they deploy the same analogical and inferential skills with online content that they learned for print.

An example of an already existing tool that could help children monitor their online reading is the Thinking Reader program, created by David Rose, Anne Meyer, and the team at the Center for Applied Special Technology (CAST). Based on universal design for learning (UDL) principles, an approach that attempts to create the most flexible, engaging ways of learning for many different kinds of children, the Thinking Reader program embeds UDL principles within text by providing different levels of strategic supports. For example, the program incorporates a hyperlink that gives background knowledge for an unknown concept or provides specific reading strategies (such as when to visualize, summarize, predict, or question), but only as needed.

The latter point is very tricky to implement. One of the consistent cautions about the use of digital technologies, particularly for struggling learners, is the tendency by children to rely too heavily on external supports, particularly when there is an option for the text to be read to them, rather than by them. Work by CAST and also a large corpus of research sponsored by the MacArthur Foundation Digital Media and Learning program illustrate how theoretically guided digital tools, particularly at the right moment in the display and with the right supports by teachers, can advance, not impede, learning. This is especially helpful for children with challenges ranging from motor and sensory impairments to dual-language learning to dyslexia.

Other tools for reading online would target more pragmatic problems such as the best uses of search engines; choosing the right search words to locate information; and, very important, learning how to evaluate information in searches, so as to see biases and attempts to influence opinion and/or consumption, and to recognize the potential for false, unsubstantiated information. Directly addressing the kind of decision-making, attention-monitoring, and executive skills that are necessary for good online reading and Internet habits is good for all learning, whatever a child's learning style and on whatever medium children use.

Within that context, discussing the good, the bad, the attractive, and the potentially harmful realities of Internet usage should become this culture's version of the sex education courses of the past and a basic part of the training that accompanies every elementary school teacher's toolbox. Julie Coiro makes the important point that we need to teach children "digital wisdom," so that they learn, first, how to make good decisions about content and, second, how to self-regulate and check their attention and ability to remember what they have read during online reading, both in and out of school.

The ultimate goal in this plan is the development of a truly biliterate brain with the capacity to allocate time and attention to deep-reading skills regardless of medium. Deep-reading skills not only provide critical antidotes to the negative effects of digital culture, like the diffusion of attention and the attrition of empathy, but also complement positive digital influences. A child who combines reading stories about refugee children with online access to actual footage of migrant refugee children waiting with their lives suspended in Greece or Turkey or upstate New York develops more empathy than one who simply reads about the situation and goes no further. At the surface level, our twenty-first-century children appear more cognizant of their connected world than ever before, but they are not necessarily building the deeper forms of knowledge about others that enable them to feel what it means to be someone else and to understand that other's feelings. As Sherry Turkle noted in *Alone Together*, our children often do better texting each other from a shared back seat than discussing their thoughts and feelings face-to-face. Deep-reading skills that involve various media may help children build a more developed compassionate imagination.

If all goes well in this proposed plan, by the time they are approximately ten to twelve years old, most children will be proficient in reading on two mediums and multiple media, and able to switch effortlessly between them for different tasks. They will have begun to learn for themselves which medium is better for which type of content and learning task, and they will know how to read deeply and think deeply, regardless of medium. If we can achieve such goals for more and more of our children, as Pope Francis wrote, society will be healthier and the world more human.

Limitations, Hurdles, and Reasons for Optimism

If we as a society want to build learning environments that promote a biliterate brain, we have to step up to the plate and deal with three large issues. First, from my perspective as a scientist, we need to invest in a great deal more research about the cognitive impacts of both print and digital mediums on all our children, particularly those with reading challenges, whether environmental or biological in origin. Second, from my perspective as an educator, we need to invest in more comprehensive professional training. Most teachers (a full 82 percent) have never been given training in the best uses of technology for children from kindergarten to grade four, much less how to teach good online reading skills to different kinds of learners. Third, from my perspective as a citizen, we must confront the access gaps that exist in our society and the world and work to eliminate them.

THE FIRST HURDLE: RESEARCH ON IMPACT

There is precious little research comparing how different mediums affect learning to read in students who have individual differences—the thousands who are failing. The seriousness of the current reality means that at the present rate, the majority of eighth-grade children could be classified as functionally illiterate in a few years' time. They will read, but they will not read proficiently or think and feel sufficiently about what they read.

The hybrid approach to the building of a biliterate brain needs to be far more carefully developed for grades K–12 with these nonproficient children in the center of our sights. This requires long-term, rigorous research, beginning with studies directly confronting the specific impacts on children's attention and memory of different mediums; the effects of the exponentially increasing time spent on digital devices with its concomitant increases in distraction; the escalating potential for addiction among our youth; and the already noted decline in empathy among young people. We need a thorough understanding of what is optimal for diverse learners at each stage of their development. We need parents, educators, and political leaders to demand such studies; we need publishers and designers to create digital innovations that are as cognitively successful as they are engaging; and we need empirical proof that this is so.

THE SECOND HURDLE: PROFESSIONAL TRAINING AND DEVELOPMENT

If two-thirds of American children are having difficulty becoming proficient readers in just one medium, what are the odds that this will work in two? Will biliteracy prove yet another class-based obstacle to their success? How can teachers be given responsibility for yet another impossible charge?

There are more reasons to be optimistic now than at any previous time. First, we know from the new research that there are six or seven basic profiles of early readers, making the identification of problem readers far easier to assess far earlier. Teachers can then better tailor their teaching to different children's needs. In the near future, digital media could change the entire trajectory of learning and teaching. For example, most children with dyslexia require up to ten times as many exposures to the letter-sound correspondence rules and common letter patterns of English than a teacher in a room of twenty-five children can easily provide, whatever the grade. For those children, the use of the digital medium would be game-changing. Think what would happen if those struggling readers could practice letter patterns and rules before other children in the class, either the day before or the same morning. Given that children with reading challenges can easily feel that something is "wrong" with them, using a digital medium in this way could provide the multiple reiterations they need and potentially display their frequently undiscovered creative strengths, thus defusing the negativity that children with dyslexia frequently, unfairly endure.

Further, there are children who will never be good screen readers and will always prefer print and vice versa. A fascinating study by Julie Coiro looked at preferences for reading by seventh-graders. Her most thought-provoking result was that the highest-performing print readers were often the lowest-performing online readers, and the converse. Whether this finding already reflects the emergence of two different reading circuits in older children today or an underlying learning difference, it is very possible that some children with dyslexia are better served by becoming digital readers early on. Certainly, the use of digital technology to provide them with maximum exposures to the sounds, grammatical functions, and meanings of the words they read in various contexts—all to be learned in their "own sweet time"—would be a great advantage for teachers and students alike.

For slightly older children for whom learning to read continues to be a struggle and for whom books have become dreaded entities, digitally interactive books and recorded audiobooks, as well as carefully chosen video games, are effective complementary media. Indeed the expanding research on video games suggests that some children's successes on these games not only increases their visual attention and eye-hand motor skills, but unobtrusively encourages learning to read when doing so is necessary to win games.

The neuroscientist and school director Gordon Sherman and his teaching staff at Newgrange School use all manner of digital tools to capture and sustain the attention of their older students with a range of learning challenges. When I visited the school, Gordon led me to the music lab, where I was presented with one of the most beautiful compositions by a young person I have ever heard, created with Garage-Band! Tapping into the intrinsic creativity of our diverse learners may be one of the greatest contributions made by educational technology today.

Nothing about this is easy, as the implementation of educational technology in US classrooms has shown. Meta-analyses of studies investigating the integrated use of various digital devices in the classroom show significant but only modest positive effects in reading, math, and science achievement for elementary and high school students when compared to traditional classrooms. This is not due to lack of interest by teachers. As noted by the publishing executive Rose Else-Mitchell, a 2017 survey on the use of educational technology indicates that two-thirds of US teachers are actively using some form of technology in their classrooms, but feel the need for more support and training.

In all likelihood, the lack of impressive results to date in the use of digital media in the classroom reflects multiple factors: the cognitive impact of the digital mediums that we are only beginning to understand; the lack of professional training and support for our teachers; and finally the great lumbering elephant in all educational research with technology: the digital access gap.

THE THIRD HURDLE: EQUAL ACCESS

If we are serious about leveling the playing field for American learners, we must square off with the complex relationship between digital access and inequality. A significant segment of American children have very few books in the home and little or no access to digital devices other than overused cell phones. According to Robert Putnam and James Heckman, the number of families in underprivileged environments is growing rapidly. They do not have the luxury of worrying whether there is too much digital exposure or too many enhanced e-books for their children. They have neither books nor computers.

These families undoubtedly represent a goodly portion of the 10,000 fourth-grade students in a study by the US Department of Education showing that children in the lower quartiles write less well on computerized tests. The report ended with the statement that "use of the computer may have widened the writing achievement gap." Children who have less exposure to books have different vocabularies and different experiences with stories and plots long familiar to other children. Children who have less exposure to digital devices and computers have a more difficult time keyboarding and much less practice using a digital medium to record their thoughts on computer-based tests—tests that many a parent and teacher as well as this author would approach with mixed feelings. If we are to fashion a code-switching reading brain for all our children, we need to figure out how to deal with both the often cited achievement gap and the less discussed digital culture gap.

In a superb report entitled "Opportunity for All?: Technology and Learning in Lower-Income Families," Victoria Rideout and Vikki Katz describe a survey of more than a thousand families in the low-to-moderate-income range. There are two different kinds of digital gaps in these families: one involves *access* to digital tools; the other, as described by the researcher Henry Jenkins, has to do with *participation*, where parents have little ability to provide either guidance or high-quality apps, leaving children to be more entertained than helped in their educational lives. This report made clear that although most of the families surveyed were connected digitally in some way, many used only cell phones, most of which were overused and over their data limits. Only 6 percent of the families had signed up for the discounted services available (in principle) to low-income families. The authors summarized their findings by saying that "access is no longer just a yes/no question. The quality of families' Internet connections, and the kinds and capabilities of devices they can access, have considerable consequences for parents and children alike."

Let me underscore: merely having access does not ensure a child's ability to use digital devices in positive ways. Susan Neuman and Donna Celano described one of the most discouraging studies to date about digital access in their report of an initiative within libraries in Philadelphia. The noble intention of the study was to investigate the effects of providing books and digital access in libraries to underserved children and families. The results ran counter to every hoped-for outcome: simply providing access to digital tools to underserved children could actually have deleterious effects, if there was no participation by parents. The children in that study did significantly worse on tests of literacy than other children did, and the disparities between groups *increased* after technological devices were introduced, particularly when the children used them for entertainment purposes.

This study highlights a pivotal and persistent mistake in the use of digital technology for education. The positive effects of digital learning cannot be reduced to issues of access or exposure. An assumption still held by many well-intentioned technology experts is that digital exposure alone will lead to huge synaptic leaps in learning, including in literacy. Such notions have their provenance in the well-meaning but ultimately overromanticized assumption that children's innate curiosity is sufficient to propel learning and literacy. Curiosity and discovery are wonderful, fruitful, and necessary but insufficient, as Neuman and Celano's work underscored. Children can learn a great deal about digital literacy without learning very much about how to become literate.

The goals of my consortium's global literacy initiative, Curious Learning, involves harnessing the curiosity of children, particularly nonliterate children in remote parts of the world, through the use of digital devices with theoretically based apps. Based on efforts to simulate the reading circuit, these apps and activities are curated or designed to foster learning to read on a digital platform at the same time that they engage the child's imagination. We have made progress toward both goals, but much more work remains to be done. We will need the concerted efforts of many groups in our country and around the globe to find solutions both to designing efficacious apps with proven results and to countering digital access gaps, particularly with regard to parental participation.

We all know that progress has never been simple for our species, even in far simpler times than ours. I am a realist and optimist and see reason to be both. On a global scale, one of the more encouraging directions involves the entrepreneur Peter Diamandis's recent XPRIZE, which will award a large monetary grant to the research team that designs digital tablets capable of increasing the reading and learning skills of children in Tanzania in both English and Swahili. If successful, this work will provide a model for many other efforts. The commitment underlying this prize and increasing numbers of other global literacy initiatives will move our world forward, if we work together across disciplinary and geographical borders.

A recent biography of Elon Musk by Ashlee Vance failed to mention that Musk is contributing significantly to the new Adult Literacy XPRIZE, but vividly pointed out that in Musk's lexicon, the word *impossible* is translated as Phase One. This letter's biliteracy proposal represents Phase One. With increased knowledge in neuroscience, education, and technology, particularly on different media and their impact, and with attention to the access gaps in our society, we will reach Phase Two: the formation of a code-switching biliterate brain that has internalized the best characteristics of both print and digital reading.

Very important, unlike my reading brain and yours, which have begun to take on the characteristics of the digital-reading mode for everything, the hope for the next generation is that they will develop distinctly different modes of reading from the outset. They will then deploy these modes automatically for different reading purposes. For example, for email they will use their faster "reading-light" mode; for more serious material, they will use a deeper-reading mode, often, perhaps, by printing out the text! If this hypothesis proves correct, there will be less of a "bleeding over" effect from whatever is the dominant mode and, more significantly, less likelihood that our children's reading brains will be short-circuited in their development. Moreover, if this hypothesis proves correct, children with the flexible medium-switching capacities of a fully biliterate brain will further the intellectual development of our species and, if Marcelo and Carola Suárez-Orozco are correct, will expand our empathic, perspective-taking abilities, too. "Our only world" would be twice blessed.

arcia/tl (attend, remember, connect, infer,
analyze/then LEAP!)

This letter's biliteracy blueprint helps us envision how we enable children to get to the other side of the cultural divide. It begins and ends with conferring on our young each of the processes that characterize deep reading in any medium: arcia/tl. This acronym represents my only half-facetious antidote to the tl; dr phenomenon that characterizes the reading of too many of our youth today. I want to reclaim and rechannel their capacities from attention to insight.

As the short-story writer Patricia McKillip wrote, "The future—any future—was simply one step at a time out of the heart." So it has been for me with the thoughts in these last three letters about the future of our children: staking out what we must preserve in the present reading brain lest we lose something irreplaceable; pointing out traps to avoid in digital media's collateral effects on the young and old; and naming some of society's gaps, in particular, digital and print access and the role of parents. Together these thoughts provide an unfinished portrait of the digital dilemmas confronting us all and point to an exciting, complex future for our children and for us. Like Flannery O'Connor, "I can, with one eye squinted, take it all as a blessing."

Whatever promise that future holds, however, we would be the most ignorant of humans not to comprehend what we the expert readers of this moment possess. The future—any future—depends on our understanding the true value of the good reader and the role of deep reading in how we live out our lives.

Kindest thoughts,
Maryanne Wolf

READER, COME HOME

> To read, we need a certain kind of silence . . . that seems increasingly elusive in our over-networked society . . . and it is not contemplation we desire but an odd sort of distraction, distraction masquerading as being in the know. In such a landscape, knowledge can't help but fall prey to illusion, albeit an illusion that is deeply seductive, with its promise that *speed can lead us to illumination*, that it is more important to react than to think deeply. . . . Reading is an act of contemplation . . . an act of resistance in a landscape of distraction. . . . It returns us to *a reckoning with time*.
> —David Ulin

> Past a certain scale there is no dissent from a technological choice. . . . What [therefore] can turn us . . . back into the sphere of our being, that joins us to our home, to each other and to other creatures . . . ? I think it is love . . . [a] particular love . . . requiring stands and acts. . . . And it implies a responsibility . . . rising out of generosity. I think that this sort of love defines the effective range of human intelligence.
> —Wendell Berry

My dear Reader,

When I was very young, I thought that "good reader" meant that one could read all the books that filled two tiny shelves in the back of a two-room schoolhouse. When I began to study in places where books were so many that they filled multiple library buildings with levels deep under the ground, I thought that "good reader" must mean reading as many of those books as possible and making their knowledge one's own. When I was a young teacher in a place whose teachers had long left, my only thought was that if I could not help those children become "good readers," they would never leave the borders of their family's indentured lives. When I first became a researcher, I chafed when studies would compare "good readers" with children with dyslexia, who worked harder than almost anyone else to understand a text. Finally, when I studied what the brain does when it retrieves the meanings of words, I learned that every meaning I possessed for "good reader" would be activated whenever I think of it.

I have added a new meaning. As discussed earlier, in *The Nicomachean Ethics* Aristotle wrote that a good society has three lives: the life of knowledge and productivity; the life of entertainment within the Greeks' special understanding of *leisure*; and finally, the life of contemplation. So, too, the "good reader."

There is the first life of the good reader in gathering information and acquiring knowledge. We are awash in this life.

There is the second life, in which reading's varied forms of entertainment are to be found in abundance: the sheer distraction and exquisite pleasure of immersion—in stories of other lives; in articles about mysterious, newly discovered exoplanets; in poems that steal our breath away. Whether we choose to escape in bodice-ripping romances; enter the painstakingly re-created worlds in novels by Kazuo Ishiguro, Abraham Verghese, or Elena Ferrante; exercise our wits in mysteries by John Irving or in biographies of saints by G. K. Chesterton or of presidents by Doris Kearns Goodwin; or discover our species' epic genetic journey with Siddhartha Mukherjee or Yuval Noah Harari, we read to take this most economic transport away from our frantically pursued everyday lives.

The third life of the good reader is the culmination of reading and the terminus of the other two lives: the reflective life, in which—whatever genre we are reading—we enter a totally invisible, personal realm, our private "holding ground" where we can contemplate all manner of human existence and ponder a universe whose real mysteries dwarf any of our imagination.

John Dunne wrote that our culture fully embodies Aristotle's first two lives of a good society, but recedes each day from the third, the contemplative one. So, too, I think, the third life of the good reader.

Years ago the philosopher Martin Heidegger felt that the great danger in an age of technological ingenuity like ours is that it could spawn an "indifference toward meditative thinking. . . . Then man would have denied and thrown away his own special nature—that he is a meditative being. Therefore, the issue is the saving of man's essential nature—the keeping of the meditative thinking alive." There is no shortage of contemporary observers of our digital culture who worry like Heidegger that the meditative dimension in human beings is threatened—by an overwhelming emphasis on materialism and consumerism, by a fractured relationship with time. As Teddy Wayne wrote in the *New York Times*: "Digital media trains us to be high-bandwidth consumers rather than meditative thinkers. We download or stream a song, article, book or movie instantly, get through it (if we're not waylaid by the infinite inventory also offered) and advance to the next immaterial thing." Or as Steve Wasserman asked in Truthdig, "Does the ethos of acceleration prized by the Internet diminish our capacity for deliberation and enfeeble our capacity for genuine reflection? Does the daily avalanche of information banish the space needed for actual wisdom? . . . Readers know . . . in their bones something we forget at our peril: that without books—indeed without literacy—the good society vanishes and barbarism triumphs."

If we are to evaluate the truth in such descriptions of a digital culture, we must examine ourselves without a cognitive flinch and look at who we are now, both as readers and as cohabitants of a shared planet. Many changes in our thinking owe as much to our biological reflex to attend to novel stimuli to survive as to a culture that floods us with continuous stimuli with our collusion. It will be what we do next with our growing consciousness of these changes that matters; whether we exacerbate the negative changes by ignoring them or redress them with increased knowledge will depend in part on what all of us do next.

It is easy to forget that the contemplative dimension that resides within us is not a given and requires intention and time to be sustained. How we reckon with the time we are given—in milliseconds, hours, and days—may well be the most important thing any of us chooses in an age of continuous flux. In her beautiful essay "Time," Eva Hoffman beseeches us to consider how "the need for reflection, for making sense of our transient condition, is time's paradoxical gift to us, and possibly the best consolation."

Hoffman's plea was unexpectedly brought home to me in a recent interview by Charlie Rose with Warren Buffett and Bill Gates. When asked what Buffett had taught him, Gates gently remarked that Buffett had taught him to "fill his calendar with spaces." In a surprising gesture Buffett pulled out a small paper calendar, less than the size of his hand, and showed all the empty spaces, quietly saying "Time is the one thing no one can buy." No one spoke for a second and the camera did not move from that avuncular face, as if to preserve in film that simplest but most difficult-to-sustain insight.

Whether we are able to attend to our capacity for reflection in this epoch is a matter of personal choice, with critical implications for us as both individuals and citizens. John Dunne saw the loss of this dimension as related to the rise of violence and conflict. I regard its gradual loss more as an outcome of our milieu's unforeseen sequelae—the constant need for efficiency: "buying time" without knowing for what purpose; decreasing attention spans, pushed beyond their cognitive limits by a flotsam of distractions and information that will never become knowledge; and the increasingly manipulated and superficial uses of knowledge that will never become wisdom.

In the first half of the twentieth century, T. S. Eliot wrote in "Choruses from 'The Rock,'" "Where is the wisdom we have lost in knowledge? Where is the knowledge we have lost in information?" In the first quarter of our century we daily conflate information with knowledge and knowledge with wisdom—with the resulting diminution of all three. Exemplified by the interactive dynamic that governs our deep-reading processes, only the allocation of time to our inferential and critical analytical functions can transform the information we read into knowledge that can be consolidated in our memory. Only this internalized knowledge, in turn, will allow us to draw analogies with and inferences from new information. The discernment of the truth and value of new information depend on this allocation of time. But the rewards are many, including, paradoxically, time itself—for uses that could otherwise go by the wayside of our lives without notice, my segue to turn to the invisible harvests that spool from the third, contemplative life.

The Contemplative Life

TIME FOR JOY

No one can visually follow all that occurs during the last nanoseconds when we read; it is beyond the current limits of brain-imaging methods. I want to follow less visible tracks with you that lead into the third life of the reader, in which we consciously perceive time in different ways, beginning with joy.

During these last moments together, therefore, I ask that you try on what Calvino described as a "rhythm of time that passes with no other aim than to let feelings and thoughts settle down, mature, and shed all impatience or ephemeral contingency." He used the Latin expression *festina lente*, which translates as "hurry slowly" or "hurry up slowly," to underscore the writer's need to slow time. I use it here to help you experience the third life more consciously: knowing how to quiet the eye and allow your thoughts to settle and be still, poised for what will follow.

I want children to learn the capacity for such cognitive patience, and I ask you now to reclaim what you may have lost. *Festina lente* gives you a release from the reduced ways most of us now read: fast if you can, slowly if you must. To possess cognitive patience is to recover a rhythm of time that allows you to attend with consciousness and intention. You read quickly (*festina*), till you are conscious (*lente*) of the thoughts to comprehend, the beauty to appreciate, the questions to remember, and, when fortunate, the insights to unfold.

From this perspective *festina lente* provides two metaphors for all the thoughts in this book about the changes in reading. At a macro level, it directs us to how we might move across the transition to a digital culture: let us hasten to meet that future, but examine it slowly with our best thoughts at our side. At a micro level, it is a metaphor for the entire arc of the good reader's reading circuit: we decode automatically until perception becomes transformed into concepts, when time becomes consciously slowed, and our whole self becomes suffused by the mental cascade where thought and feeling converge. We may hurry to enter that inside place, but let us relearn how to stop, still our lives, and only leave that home for the self in our own time.

I have been very frugal with the use of the word *self.* But we come now to the core of the third reading life, the home where both the self and, perhaps, the soul lie side by side, and where we can look about us more knowingly through the lenses of others' thoughts. There are few better attempts to portray this invisible habitat of the reader's interior self than Virginia Woolf's description of Mrs. Ramsay in *To the Lighthouse.* As she reads poems by Shakespeare, Mrs. Ramsay begins to connect her insights into the sonnets to the whole of her life and the life of her family. Her entire being is suffused with waves of new understanding and fresh feelings of joy, while her husband looks on with that peculiar condescension that is the long-laid consequence of circumscribing the one we love and that renders the viewer oblivious to the maelstrom of thoughts and feelings the other has entered into unnoted.

To you who like Mrs. Ramsay know the place we enter when we leave the surface of our self behind and are released from time, there is a suspended joy with little parallel. Such joy is no random event reached by serendipity or a temperament disposed to happiness; rather, it is the perquisite of the hard-won thoughts and feelings of the person who makes room and time for it.

Few historical individuals better illumine the life-altering importance of the joy that reading imparts, even in the direst of circumstances, than Dietrich Bonhoeffer. Described earlier in Letter Three, Bonhoeffer wrote one of the most moving books I have ever read, *Letters and Papers from Prison*, after being thrown into concentration camps for his views about Nazi Germany. The letters portray an embattled, unyielding spirit kept alive in very large part by what he could read—to himself (the one luxury his illustrious family could arrange to send him), to his fellow prisoners, and, as revealing of his nature as his writing, to his prison guards.

Most striking in his letters is the unalloyed happiness that Bonhoeffer gained from all that he read, which he then passed on to others despite his own very deep despair. In one letter to his young fiancée, he wrote, "Your prayers and kind thoughts, passages from the Bible . . . pieces of music, books—all are invested with life and reality as never before. I live in a great, unseen realm of whose real existence I'm in no doubt." I believe that it was the unseen sanctuary within the reading act that sustained him through every deprivation till the end.

When he left Buchenwald for Flossenbürg, where he was executed just days before the American liberation and Adolf Hitler's suicide, Bonhoeffer took the Bible, Goethe, and Plutarch to accompany him. These, the books of his faith in God and the symbols of his enduring hope in the deepest good in human life and in nature, preserved him till he died. In the words of another prisoner, a British intelligence officer, "he always seemed to me to diffuse an atmosphere of happiness, of joy in every smallest event in life. . . . He was one of the very few men that I have met to whom his God was real and ever close to him. . . . He was, without exception, the finest and most lovable man I have ever met." My hope for my children and my children's children and yours is that they, like Bonhoeffer, will know where to find the many forms of joy that reside in the secret holding places in the reading life and the sanctuary it gives each of us who seeks it.

An unexpected, contemporary example of the powerful nature of this dimension in the reading act surprised me just a little while ago. The philosopher Bernard Stiegler, director of the Institut de recherche et d'innovation at the Pompidou Museum in Paris, invited me to present my research at a conference. It was a nerve-racking event for me that ended with a dinner afterward, attended by no fewer than fifteen men and myself, regrettably both the only non–French speaker and the only female. Seated next to Professor Stiegler and determined not to betray my shyness in the situation, I grasped at conversational straws and asked him how he had come to be a philosopher. After a slight, but notable pause, he said, "In prison." After an equally slight pause, one that I hoped would convey my attempt at politeness, I asked the impossible-to-repress question "But why?" To which he replied, "Armed robbery. I was in prison quite a few years."

I blurted out the most immediate hypothesis: "You were political . . . part of the French Red Brigade?" That was the beginning of the dialogue that Professor Stiegler and I began about what happens in the life of a person who is imprisoned, in this case both for conscience and for crime. Not unlike what Nelson Mandela related in *Long Walk to Freedom*, or Malcolm X in his autobiography, Stiegler read first for escape from his prison reality and then for what became an almost insatiable desire to learn. He discovered philosophy from books that a group of volunteers managed to bring him weekly, similar to the selfless work by the Reader Organisation in Great Britain. By his last year in prison, he was reading ten to twelve hours a day with what he described as "a contentment and joy unparalleled" in his life, either before or after.

The rest of the story is the stuff of Parisian legend. The eminent French philosopher Jacques Derrida asked to meet Stiegler upon his release. After their meeting, Stiegler reentered university, completed his dissertation with Derrida, and became one of the most thought-provoking, albeit controversial philosophers in France. His lifework became a series of efforts to give new perspective to how humans can live meaningful lives in a technological culture. Elaborated elsewhere, his evocative concept of *pharmakon*, a "remedy that contains a poison with a therapeutic virtue," has helped to sharpen my own perspective on technology's complex contributions to society. But it was not only his difficult dialectical contributions to modern thought that I left Paris with, but rather his living example of the contributions that reading makes—both to sustaining the self across adversity and to redirecting thoughts beyond the self to the good of others.

TIME FOR SOCIAL GOOD

> We are so distracted by and engulfed by the technologies we've created and by the constant barrage of so-called information that comes our way, that more than ever to immerse yourself in an involving book seems socially useful. . . . The place of stillness that you have to go to write, but also to read seriously, is the point where you can actually make responsible decisions, where you can actually engage productively with an otherwise scary and unmanageable world. . . .
>
> —Jonathan Franzen

Bonhoeffer and Stiegler are two examples of human beings in whom the third reading life supports the self through otherwise impossible circumstances and becomes the basis of emulable service to others. The "place of stillness" that Jonathan Franzen describes is the reflective domain in which the act of reading allows us to think critically for ourselves and to make responsible decisions, which in the process become socially useful acts.

In a recent essay about our values as a nation, Marilynne Robinson wrote, "I do believe that we stand at a threshold, as Bonhoeffer did, and that the example of his life obliges me to speak about the gravity of our historical moment as I see it, in the knowledge that no society is at any time immune to moral catastrophe. . . . We owe it to him to acknowledge a bitter lesson he learned before us, that these challenges can be understood too late."

We live in a historical "hinge moment" as Robert Darnton called it, en route to whole new forms of communication, cognition, and choices that are ultimately, deeply ethical. Unlike during other great transitions, we have the science, the technology, and the ethical imagination necessary to understand the challenges we face before it is too late—if we choose to do so. As depicted earlier, we need to confront the reality that when bombarded with too many options, our default can be to rely on information that places few demands upon thinking. More and more of us would then think we know something based on information whose source was chosen because it conforms to how and what we thought before. Thus, though we are seemingly well armed, there begins to be less and less motivation to think more deeply, much less try on views that differ from one's own. We think we know enough, that misleading mental state that lulls us into a form of passive cognitive complacency that precludes further reflection and opens wide the door for others to think for us.

This is a long-known recipe for intellectual, social, and moral neglect and the fraying of societal order. At stake here is the ultimate message of this book: that any version of the digital chain hypothesis, strong or weak, poses threats to the use of our most reflective capacities if we remain unaware of this potential, with profound implications for the future of a democratic society. The atrophy and gradual disuse of our analytical and reflective capacities as individuals are the worst enemies of a truly democratic society, for whatever reason, in whatever medium, in whatever age.

Twenty years ago Martha Nussbaum wrote about the susceptibility and the decision making of citizens who have ceded their thinking to others:

> It would be catastrophic to become a nation of technically competent people who have lost the ability to think critically, to examine themselves, and to respect the humanity and diversity of others. And yet, unless we support these endeavors, it is in such a nation that we may well live. It is therefore very urgent right now to support curricular efforts aimed at producing citizens who can take charge of their own reasoning, who can see the different and foreign not as a threat to be resisted but as an invitation to explore and understand, expanding their own minds and their capacity for citizenship.

Nussbaum's plea for a more thoughtful, compassionate, diverse citizenry could not be more urgent or timely. If we gradually lose the ability to examine how we think, we will also lose the ability to examine dispassionately how those who would govern us think. The worst atrocities of the twentieth century bear tragic witness to what occurs when a society fails to examine its own actions and cedes its analytical powers to those who tell them how to think and what to fear. Bonhoeffer described this old scenario from his prison cell:

If we look more closely, we see that any violent display of power, whether political or religious, produces an outburst of folly in a large part of mankind;
indeed, this seems actually to be a psychological
and sociological law: the power of some needs the
folly of the others. It is not that certain human capacities, intellectual capacities for instance, become
stunted or destroyed, but rather that the upsurge
of power makes such an overwhelming impression
that men are deprived of their independent judgment
and . . . give up trying to assess the new state of af-
fairs for themselves.

Two of the greatest mistakes of the twenty-first century,
therefore, would be to ignore those of the twentieth century
and fail to evaluate whether we have already begun to cede
our critical analytical powers and independent judgment
to others in our increasingly fissured society. Few people,
if pressed, would contest that such a diminishment of our
collective critical faculties has already begun. What would
be contested is in whom and why.

I could never have imagined that research about the
changes in the reading brain, most of which reflect increasing adaptations to a digital culture, would have implications for a democratic society. Yet that is my conclusion.
In a dialogue between Umberto Eco and Cardinal Carlo
Maria Martini, the cardinal reiterated a timeless view of
the democratic process that is pertinent to this conclusion:
"The delicate game of democracy provides for a dialectic
of opinions and beliefs in the hope that such exchange will
expand the collective moral conscience that is the basis of
orderly cohabitation."

The most important contribution of the invention of
written language to the species is a democratic foundation

for critical, inferential reasoning and reflective capacities. This is the basis of a collective conscience. If we in the twenty-first century are to preserve a vital collective conscience, we must ensure that all members of our society are able to read and think both deeply and well. We will fail as a society if we do not educate our children and reeducate all of our citizenry to the responsibility of each citizen to process information vigilantly, critically, and wisely across media. And we will fail as a society as surely as societies of the twentieth century if we do not recognize and acknowledge the capacity for reflective reasoning in those who disagree with us.

As Nadine Strossen writes persuasively in her new book, *Hate: Why We Should Resist It with Free Speech, Not Censorship*, a democracy succeeds only when the rights, thoughts, and aspirations of all its citizens are respected and given voice and its citizens believe that this is true, regardless of their viewpoint. The great, insufficiently discussed danger to a democracy stems not from the expression of different views but from the failure to ensure that all citizens are educated to use their full intellectual powers in forming those views. The vacuum that occurs when this is not realized leads ineluctably to a vulnerability to demagoguery, where falsely raised hopes and falsely raised fears trump reason and the capacity for reflective thinking recedes, along with its influence on rational, empathic decision making.

Most people never become aware of any of this. As my recent faux experiment reading Hesse's *Magister Ludi* illustrates, a personal awareness of the gradual disuse of our reflective faculties, much less a societal awareness, is a weak and porous thing—to be tested, not trusted. Just as I worry that in their overreliance on external sources of information, our young will not know what they do not know, I worry equally that we, their guides, do not realize the insidious narrowing of our own thinking, the imperceptible shortening of our attention to complex issues, the unsuspected diminishing of our ability to write, read, or think past 140 characters. We must all take stock of who we are as readers, writers, and thinkers.

The good readers of a society are both its canaries—which detect the presence of danger to its members—and its guardians of our common humanity. The final perquisite of the third reading life is the ability to transform information into knowledge and knowledge into wisdom. Indeed, just as Margaret Levi has suggested for the basis of altruism, the combining of our highest intellectual and empathic powers with our capacity for virtue may well be why our species has continued. If these capacities are endangered, if good readers are endangered, so are we all. If they are supported, we will have not only an antidote to the weaknesses of a digital culture but a key to propelling our culture's greatest potential into the future: wise action.

TIME FOR WISDOM

> Wisdom, I conclude, is not contemplation alone, not action alone, but contemplation in action.
> —John Dunne

Of all the gifts that the third life of the good reader bestows, wisdom, the highest form of cognition, is its ultimate expression. In *The Language Animal*, the philosopher Charles Taylor began a luminous passage about language with a quote from Wilhelm von Humboldt that brings to life the human "drive to articulate," which underlies a search for wisdom: "There is always a 'feeling that there is something which the language does not directly contain, but which the [mind/soul], spurred on by language, must supply; and the [drive], in turn, to couple everything felt by the soul with a sound.'" In Taylor's view, the very nature of "possessing a language is to be continuously involved in trying to extend its powers of articulation."

So also the experience in the third life of the good reader: to be continuously engaged in trying to reach and express our best thoughts so as to expand an ever truer, more beautiful understanding of the universe and to lead lives based on this vision. To embark on such a quest is the furthest goal of deep reading and the beginning of wisdom, but not its end. Just as Proust articulated years ago, "the end of [the author's] wisdom is but the beginning of ours." For some years now, these words have been my aide-mémoire for knowing when to stop and prepare the good reader—you, my dear reader—to take over the work that lies ahead of us all.

The Future of Reading and Good Readers

> Word-work is sublime . . . because it is generative; it makes meaning that secures our difference, our human difference— the way in which we are like no other life. We die. That may be the meaning of life. But we do language. That may be the measure of our lives.
>
> —Toni Morrison

From the first letter to the last, these pages celebrate the human-driven achievement that is the reading brain. In between its pages, my hope was to engage in dialogue with you the reader about my concerns. First, will the very plasticity of a reading brain that reflects the characteristics of digital media precipitate the atrophy of our most essential thought processes—critical analysis, empathy, and reflection—to the detriment of our democratic society? Second, will the formation of these same processes be threatened in our young? To be sure, each of these human processes is perennially endangered. Yet each has accelerated across the centuries. We can take comfort from that.

Less comfort is to be found in my third concern, because it is equally beneficial to our development. We humans appear to be born with an unassuageable drive to add to our capacities and go beyond our perceived limits. When we cannot, we create new tools and technologies that do so for us. Indeed, the very plasticity of the human brain primes and permits us to do this. But the plasticity in the brain also has a wisdom of its own about altering some capacities (such as attention and memory) when we try to gainsay our perceptual and intellectual limits with technology's new tools. Just as there were "misses" in evolution, in which whole species, traits, or abilities vanished because the environment did not support their continuation, there can be misses in the epigenetic changes to our cognitive capacities as we enthusiastically acquire newly essential skills that prepare us for a future whose parameters we can barely imagine.

This is the digital dilemma that is being acted out this moment in the cognitive, affective, and ethical processes now connected in the present reading circuitry and now threatened. How easy it would be to short-circuit these processes, which have made us who we have been as readers till now. How simple it would be to leap to new modes of acquiring more knowledge more quickly and ignore the growing gaps between the information we read and the analysis and reflection we apply to it. It will be an "act of resistance," as David Ulin expressed it, to pause for a moment and examine with all our intelligence who we want to be next and what will be the best combinatoria of faculties in the reading brains of our future generations.

By now you realize that the deep-reading brain is both a real, flesh-and-cranial-bone reality and a metaphor for the continuous expansion of human intelligence and virtue. If sometimes I am too fearful about short-circuiting it in future generations, I simultaneously hope and trust in this circuit's pluripotential capacities to embody all of our species' exponentially growing intellectual, affective, and moral faculties.

This is our generation's hinge moment: the time when we decide to take the true measure of our lives. If we act wisely at this cultural, cognitive crossroads, I believe, not unlike what Charles Darwin hoped for our species' future, that we will forge ever more elaborated reading-brain circuits capable of "endless forms most beautiful."

• • •

Festina lente, dear and good reader. Come home.
Godspeed,
Maryanne

ACKNOWLEDGMENTS

"Every book has a life of its own." Those are the words my prescient HarperCollins editor, Gail Winston, told me when I finished my first book ten years ago. I think of that now because it could not be more true for this book, with all the people who contributed to its gestation and development, beginning with my mother, Mary Elizabeth Beckman Wolf. A seemingly ordinary woman, she was an extraordinary, perhaps even brilliant autodidact who never stopped reading books or nurturing all her children, grandchildren, and even great-grandchildren till the last week of her life. Two days before she died, I was able to tell her that this book would be dedicated to her, my best friend. I do not doubt that she heard me. She always did, and if I am very lucky, she still does.

My two sons, Ben Wolf Noam in the art world and David Wolf Noam at Google, sometimes don't look as though they are listening as they text and multitask, but I know they do. Their ever-wiser insights now guide me as much as I hope mine guide them. If the titles they suggested for this book (e.g., *tl;dr!*) have not been used, their many thoughts about its central themes are part of the dialogue that ran through my mind as I wrote it. I cannot love them more, and I cannot thank them enough.

The reality is that I can't thank many people enough for the multiple ways they contributed to the writing of this book. Gail Winston, my editor, and Anne Edelstein, my literary agent, were close to co-parents of this work. No one could have given me more careful and more cogent help in draft after draft. I once thought of them, *sensu* Dante, as my Beatrices; but I have come to think of them as my indispensable glia, the special cells that scaffold, heal, graft, and guide the brain's first neurons to their final home. So it has been with Anne's and Gail's support during this book's migration to its final version. If some of you find such a term rather abstruse, for me it is my highest praise for two extraordinary professional women whom I feel grateful to call friends. I am also very grateful to two other friends, Dr. Aurelio Maria Mottola, the director of the Italian publisher Vita e Pensiero, for his powerful insights into language and literature in Letters One through Four; and the playwright Cathy Tempelsman for her kind help with the title.

No book, no article, no essay by me could have been written without the years of work by my research colleagues and graduate students at the Tufts University Center for Reading and Language Research. The list always begins with my former assistant director, the child linguist Stephanie Gottwald, whose dedication to children is matched only by those she worked with at CRLR over the years, including Katharine Donnelly Adams, Maya Alivisatos, Mirit Barzillai, Surina Basho, Terry Joffe Benaryeh, Kathleen Biddle, Ellen Boiselle, Patricia Bowers, Joanna Christodoulou, Colleen Cunningham, Terry Deeney, Patrick Donnelly, Wendy Galante, Yvonne Gil, Eric Glickman-Tondreau, Anneli Hershman, Tami Katzir, Cynthia Krug, Lynne Tomer Miller, Maya Misra, Cathy Moritz, Elizabeth Norton, Beth O'Brien, Melissa Orkin, Alyssa O'Rourke, Ola Ozernov-Palchik, Catherine Stoodley, Catherine Ullman Shade, Laura Vanderburg, and many others who should be mentioned, save for length. I want to be sure in this book to thank Mirit Barzillai for her help and thoughts on technology and children; Tami Katzir and Melissa Orkin for their important new insights into fluency and affect; Ola Ozernov-Palchik for her exceptional research on reading prediction and music; and Daniela Traficante and Valentina Andolfi for their exciting work on an Italian version of the RAVE-O intervention.

In the last year, Niermala Singh-Mohan both helped coordinate the activities of the center with Drs. Gottwald and Orkin and helped prepare this manuscript for publication, for which she deserves a badge of honor! Similarly deserving, Catherine Stoodley, a prolific neuroscientist at American University, has now wonderfully illustrated three of my books with her unique, whimsical views of the reading brain. She is twice gifted.

Three other groups of colleagues have undergirded and expanded my program of research over the last years. My National Institute of Child Health and Human Development research partners and dear friends, Robin Morris and Maureen Lovett, and I have been working together for more than two decades on intervention for children with dyslexia and other reading challenges. We are singularly grateful for the enormous support of this work by the NICHD, under the directorships of Reid Lyon and Peggy McCardle. I think of Maureen and Robin as my research glia and the best colleagues one could ever have. Both of them are also involved in our newest research collaboration on global literacy (Curious Learning) along with Stephanie Gottwald (yes, she wears many hats!), Tinsley Galyean, and my MIT Media Lab colleague, the social roboticist Cynthia Breazeal, as well as Eric Glickman-Tondreau and Taylor Thompson.

Most recently, I am indebted to my friends and colleagues at UCLA Carola and Marcelo Suárez-Orozco for their critical work on social justice and children: from their ongoing research into the lives of immigrant children to our shared work on complex diverse learners. I am indebted to them, the neurologist Antonio Battro, and Monsignor Marcelo Sánchez Sorondo, the chancellor of the Pontifical Academy of Sciences, for inviting me to present my research on literacy at multiple Vatican meetings concerning the world's disenfranchised children. In related work, I want to thank my UCSF colleagues in the School of Medicine, Fumiko Hoeft and Maria Luisa Gorno-Tempini at the Dyslexia Center, for their cutting-edge neuroscience research into dyslexia and their commitment to its application in our schools. Together these colleagues across California and I hope to coordinate efforts in universities, clinics, and schools, both public and independent, to provide literacy for as many children as we can reach, especially those with reading and learning challenges.

Although they never conducted a day of research with me, my Cambridge friends gave me the kind of support that every woman writer needs: other women writers and artists. I will always be grateful to the wonderful novelists Gish Jen and Allegra Goodman, the Boston architect Maryann Thompson, and the Harvard lepidopterist Naomi Pierce (she's the one who proved Vladimir Nabokov correct in his study of butterfly migration patterns!) for their inimitable encouragement and fellowship over a hundred breakfasts. To Jacqueline Olds, there was never a better friend over just as many lunches; and to Deborah Dumaine, Lenore Dickinson, and Christine Herbes-Sommers, there were never better dinners with friends.

I could not have done the research for the present book without the magnanimous support of the Tufts University administration, particularly Dean James Glaser, Dean Joe Auner, and President Anthony Monaco. They allowed and indeed encouraged me to take two years of leave to write this book at the Stanford University Center for Advanced Study in the Behavioral Sciences (CASBS), for which I will always be grateful. My colleagues in the Eliot-Pearson Department of Child Study and Human Development and in the Cognitive Sciences program have also been a source of much support, particularly Chip Gidney, Ray Jackendoff, Fran Jacobs, Gina Kuperberg, and my chair, David Henry Feldman. My dear friend and remarkable Tufts colleague, the late Jerry Meldon, will always be missed by me and all those who knew him.

CASBS occupies a special place in the life of this book and my other books as well. Under the wise, visionary direction of Margaret Levi (see note in Letter Nine on her work on "reciprocal altruism"), CASBS provided an intellectual sanctuary for me and my fellow scholars so that we could have a moment out of time in which to write, discuss with one another across disciplinary boundaries, and, through the process, generate new directions of thought. The entire staff of CASBS—from Margaret and Associate Director Sally Schroeder to my favorite technology expert, Ravi Shivana—created an unparalleled space for reflection and its products. The life of this book began there.

And it continued in the summers in one of the most beautiful villages in the world, Talloires, France, where Tufts has its international center and summer school on the shores of Lake Annecy. Thanks to the generosity and kindness of Gabriella Goldstein, the director of the Talloires program, I have spent part of my last summers there, writing this book in the studio of the French artist Laure Tesnière. I am so grateful to both of these amazing women.

There is another amazing woman whom I thank upon every possible occasion and who with her husband, Brad, has made the last decade of my work in reading intervention and global literacy possible: Barbara Evans. She and Brad funded much of my intervention research and the training of many graduate students who have gone on to become teachers or to conduct research on literacy and dyslexia. Most of all, Barbara has been a source of kindness and inspiration to me, always supportive, always gently urging me and everyone she knows to do their best to help children everywhere. Barbara and Brad are two of the finest persons I know.

I want to end these thoughts of thanks where I began— with my mother, my family, and my friends. My mother and father were the best parents one could imagine, never ceasing to do their best to support each of their four children, Joe, Karen, Greg, and myself, in every way they knew how. I am as lucky with my siblings and their spouses, Barbara, Barry, and Jeanne, as I was with my parents. There is no coincidence there, just the best of fortune and hard work by all of us to preserve the physical, moral, and spiritual legacy of Frank and Mary Wolf.

I feel the same for my dearest living friends: my sister Karen, Heidi and Thomas Bally, Cinthia Coletti Steward, Christine Herbes-Sommers, Sigi Rotmensch, Aurelio Maria Mottola, and Lotte Noam, and for those who have died, Ulli Kesper Grossman, Ken Sokoloff, David Swinney, Tammy Unger, and Father John S. Dunne, my teacher and friend, whose work has accompanied my thoughts throughout this book.

I am grateful to you all. I could never have written this book without each of you. This is the underlying meaning of "Every book has a life of its own."

CREDITS

NOTES

EPIGRAPH

vi **"We are in a different phase"**: J. Enriquez and S. Gullans, *Evolving Ourselves: How Unnatural Selection and Nonrandom Mutation Are Changing Life on Earth* (New York: Current, 2017), 180, 259.

vii **Reading is an act**: D. L. Ulin, *The Lost Art of Reading: Why Books Matter in a Distracted Time* (Seattle, WA: Sasquatch Books, 2010), 150.

LETTER ONE:
READING, THE CANARY IN THE MIND

1 **"Fielding calls out"**: B. Collins, "Dear Reader," in *The Art of Drowning* (Pittsburgh: University of Pittsburgh Press, 1995), 3.

1 **galactic changes**: I refer both to work of futurists such as Enriquez and Gullans, *Evolving Ourselves: How Unnatural Selection and Nonrandom Mutation Are Changing Life on Earth* and also to a new study by astrophysicists at Northwestern University that now indicates that each of us contains the stuff (atoms of carbon, nitrogen, oxygen, etc.) of not only our own galaxy but also other galaxies. See *Monthly Notices of the Royal Astronomical Society*, July 26, 2017.

1 ***human beings were never***: This is the beginning of my book *Proust and the Squid: The Story and Science of the Reading Brain* (New York: HarperCollins, 2007).

4 ***Duino Elegies***: R. M. Rilke, *Duineser Elegien*, trans. A. Poulin, Jr. (Boston: Houghton Mifflin, 1977).

4 **Peace Corps–like stint in rural Hawaii**: This is the project sponsored by University of Notre Dame in the CILA program. Eric Ward and I and Henry and Tony Lemoine volunteered to be teachers in a school in Waialua, Hawaii, where there were no longer enough teachers for the children and where most of the parents had come from the Philippine Islands to work on the sugar plantation there.

6 ***Proust and the Squid***: See Wolf, *Proust and the Squid*.

7 **Steven Hirsh**: Professor of classics, Tufts University, to whom I continue to be grateful for his almost yearlong tutorial on Socrates and Plato.

7 **Walter Ong:** W. Ong, *Orality and Literacy* (London: Methuen, 1982).
8 **part of deep reading:** First used by Sven Birkerts in *Gutenberg Elegies* and used more specifically (cognitively) by me in my research. See M. Wolf and M. Barzillai, "The Importance of Deep Reading," *Educational Leadership* 66, no. 6 (2009): 32–37. I am indebted to Nicholas Carr for his general incorporation of the term in his book aptly called *The Shallows*.
9 **we have choices to make:** Enriquez and Gullans, *Evolving Ourselves*.
10 **"fertile miracle of communication":** M. Proust, *On Reading*, ed. J. Autret, trans. W. Burford (New York: Macmillan, 1971; originally published 1906), 31.
10 *Letters to a Young Poet*: R. M. Rilke, *Letters to a Young Poet*, trans. M. D. H. Norton (New York: W. W. Norton, 1954). See also Rilke, *Briefe an einen jungen Dichter* (Wiesbaden: Insel-Verlag, 1952). These letters were exchanged with Franz Xaver Kappus between 1902 and 1908.
10 *Six Memos for the Next Millennium*: I. Calvino, *Six Memos for the Next Millennium* (Cambridge, MA: Harvard University Press, 1988).
12 **Kant's three questions:** See J. S. Dunne, *Love's Mind: An Essay on Contemplative Life* (Notre Dame, IN: University of Notre Dame Press, 1993).
12 **global literacy:** See the work that my colleagues at Curious Learning: A Global Literacy Project are pursuing in the last chapter of M. Wolf, *Tales of Literacy for the 21st Century* (Oxford, UK: Oxford University Press, 2016). This work has been presented at four meetings of the Pontifical Academy of Sciences in Vatican City. Chapters include M. Wolf et al., "The Reading Brain, Global Literacy, and the Eradication of Poverty," *Proceedings of Bread and Brain, Education and Poverty* (Vatican City: Pontifical Academy of Social Sciences, 2014); M. Wolf et al., "Global Literacy and Socially Excluded Peoples," *Proceedings of The Emergency of the Socially Excluded* (Vatican City: Pontifical Academy of Social Sciences, 2013).
13 **Aristotle wrote:** See Dunne, *Love's Mind*.
14 **"iron sharpens iron":** J. Pieper, *The Silence of St. Thomas*, trans. John Murray and Daniel O'Connor (South Bend, IN: St. Augustine's Press, 1957), 5.
14 **"It seemed to me":** Proust is quoted in this translation in M. Edmundson, *Why Reading?* (New York: Bloomsbury, 2004), 4.

LETTER TWO:
UNDER THE BIG TOP

15 **"The Brain—is wider than the Sky":** E. Dickinson, *The Complete Poems of Emily Dickinson*, ed. T. J. Johnson (Boston: Little, Brown, 1961). Wikisource, 6320.

16 "Tell all the truth, but tell it slant": Ibid. Wikisource, 1129.

16 "connected to one another": D. Eagleman, *Incognito: The Secret Lives of the Brain* (New York: Viking, 2011), 1.

16 circuit for reading: This letter is heavily based on research summarized in "A Neuroscientist's Tale of Words," chap. 4 of M. Wolf, *Tales of Literacy for the 21st Century* (Oxford, UK: Oxford University Press, 2016). See work on the concept of circuits in S. Petersen and W. Singer, "Macrocircuits," *Current Opinion in Neurobiology* 23, no. 2 (2013): 159–61. See the important work on reading circuits by B. A. Wandell and J. D. Yeatman, "Biological Development of Reading Circuits," *Current Opinion in Neurobiology* 23, no. 2 (2013): 261–68; B. L. Schlaggar and B. D. McCandliss, "Development of Neural Systems for Reading," *Annual Review of Neuroscience* 30 (2007): 475–503; J. Grainger and P. J. Holcomb, "'Watching the Word Go By': On the Time-course of Component Processes in Visual Word Recognition," *Language and Linguistics Compass* 3, no. 1 (2009): 128–56.

17 the brain recycles: The term *neuronal recycling* has been used by Stanislas Dehaene to refer to "the partial or total invasion of a cortical territory initially devoted to a different function, by a cultural invention. . . . Neuronal recycling is also a form of reorientation or retraining: it transforms an ancient function . . . into a novel function that is more useful in the present cultural context." S. Dehaene, *Reading in the Brain: The New Science of How We Read* (New York: Viking, 2009), 147.

18 Chinese, character-based reading-brain circuit: See D. J. Bolger, C. A. Perfetti, and W. Schneider, "Cross-Cultural Effects on the Brain Revisited: Universal Structures plus Writing System Variation," *Human Brain Mapping* 25, no. 1 (May 2005): 92–104.

18 This is not the case: I will leave aside for now a discussion of outliers such as Jean-Paul Sartre and the novelist Penelope Fitzgerald who seemingly developed this capacity by themselves before they could barely speak. See discussion in M. Wolf, *Proust and the Squid: The Story and Science of the Reading Brain* (New York: HarperCollins, 2007).

18 *neuroplasticity*: See discussion in M. Wolf, *Tales of Literacy in the 21st Century*.

19 Donald Hebb: Originally published in 1949 and reprinted as D. Hebb, *The Organization of Behavior: A Neuropsychological Theory* (Mahwah, NJ: Psychology Press, 2002).

22 Figure 1: Catherine Stoodley's overview is based on several meta-analyses of brain imaging studies of reading. See in particular A. Martin, M. Schurz, M. Kronbichler, and F. Richlan, "Reading in the Brain of Children and Adults: A Meta-Analysis of 40 Functional Magnetic Resonance Imaging Studies," *Human Brain Mapping* 36, no. 5 (May 2015): 1963–81; Grainger and Holcomb, "'Watching the Word Go By.'"

22 **Chinese and Japanese Kanji:** Bolger, Perfetti, and Schneider, "Cross-Cultural Effects on the Brain Revisited."

23 **biological spotlights:** See the work on attention by Earl Miller and Timothy Buschman, e.g., E. K. Miller and T. J. Buschman, "Cortical Circuits for the Control of Attention," *Current Opinion in Neurobiology* 23, no. 2 (April 2013): 216–22.

23 **orienting attentional system:** For a fuller description of attention, memory, and visual systems in reading, see Wolf, *Proust and the Squid,* and Wolf, *Tales of Literacy for the 21st Century.*

26 *retinotopic organization:* For comprehensive descriptions of the role of the visual system in reading, see B. A. Wandell, "The Neurobiological Basis of Seeing Words," *Annals of the New York Academy of Sciences* 1224, no. 1 (April 2011): 63–80; Wandell and Yeatman, "Biological Development of Reading Circuits."

26 *representation:* See B. A. Wandell, A. M. Rauschecker, and J. D. Yeatman, "Learning to See Words," *Annual Review of Psychology* 63 (2012): 31–53.

28 **Even if we just imagine:** The work on visual representations has been very influenced by the program of research by Stephen Kosslyn, with a landmark study: S. M. Kosslyn, N. M. Alpert, W. L. Thompson, et al., "Visual Mental Imagery Activates Topographically Organized Visual Cortex: PET Investigations," *Journal of Cognitive Neuroscience* 5, no. 3 (Summer 1993): 263–87.

29 **where the occipital and temporal lobes meet:** This somewhat controversial area is what Dehaene, Cohen, and McCandliss, among others, call the Visual Word Form Area, or VWFA. Dehaene also refers to it as the Letterbox. Others label this region differently: e.g., Yale University's Ken Pugh refers to it simply as the occipital-temporal junction. British researchers such as Cathy Price conceptualize the area more broadly as a convergence zone with polymodal interactions among visual, auditory, and tactile areas and with involvement in various functions such as word retrieval. See C. J. Price and J. T. Devlin, "The Myth of the Visual Word Form Area," *Neuroimage* 19, no. 3 (July 2003): 473–81.

29 **forty-four different phonemes:** The largest body of research on reading over the last four decades has emphasized the essential role that phonemes and their underlying phonological processes play in the acquisition of the alphabetic code and in reading challenges such as dyslexia. See the excellent recent summary in M. Seidenberg, *Language at the Speed of Sight: How We Read, Why So Many Can't, and What Can Be Done About It* (New York: Basic Books, 2017).

29 **based on probabilities and prediction:** See the important work by Andy Clark on how prediction prepares perception, e.g., A. Clark, "Whatever Next? Predictive Brains, Situated Agents, and the Future of Cognitive Science," *Behavioral and Brain Sciences* 36, no. 3 (June 2013): 181–204. Using multiple forms of imaging in her re-

search, Gina Kuperberg shows that such predictions are at work in everything from identifying a letter to selecting the most predictable meaning of a word. Thus what we know speeds up the recognition of what we see. See G. R. Kuperberg and T. F. Jaeger, "What Do We Mean by Prediction in Language Comprehension?," *Language and Cognitive Neuroscience* 31, no. 1 (2016): 32–59.

30 **all manner of interesting:** See the early work in priming research by the cognitive scientist David Swinney on how we activate the multiple meanings of words unconsciously whenever we see the word displayed; see, e.g., D. A. Swinney and D. T. Hakes, "Effects of Prior Context upon Lexical Access During Sentence Comprehension," *Journal of Verbal Learning and Verbal Behavior* 15, no. 6 (December 1976): 681–89.

30 **physically act it out:** Fascinating research suggests how the motoric system is activated when we initially encounter the word in a text. For activation for reading verbs, see in particular F. Pulvermüller, "Brain Mechanisms Linking Language and Action," *Nature Reviews Neuroscience* 6, no. 7 (July 2005): 279–95. See also Raymond Mar's work on embodied comprehension, e.g., H. M. Chow, R. A. Mar, Y. Xu, et al., "Embodied Comprehension of Stories: Interactions Between Language Regions and Modality-Specific Mechanisms," *Journal of Cognitive Neuroscience* 26, no. 2 (February 2014): 279–95.

30 **"semantic neighborhood":** For an excellent, accessible summary of work on how semantic processes work, see R. Jackendoff, *A User's Guide to Thought and Meaning* (New York: Oxford University Press, 2012).

31 **Anna Karenina:** L. Tolstoy, *Anna Karenina*, trans. Constance Garnett (New York: Barnes and Noble Classics, 1973; originally published 1877).

31 *angular gyrus:* This region plays an integrative role during the acquisition of reading. Earlier work by the behavioral neurologist Norman Geschwind placed the angular gyrus in a more central role in his early models of reading. Current imaging studies show its activation in semantic processing, particularly when monitoring the linking of semantic and phonological information. See, e.g., Kuperberg and Jaeger, "What Do We Mean by Prediction in Language Comprehension?," and research by Mark Seidenberg and his colleagues, e.g., W. W. Graves, J. R. Binder, R. H. Desai, et al., "Anatomy Is Strategy: Skilled Reading Differences Associated with Structural Connectivity Differences in the Reading Network," *Brain and Language* 133 (June 2014): 1–13.

33 **our words contain and momentarily activate:** See Swinney and Hakes, "Effects of Prior Context upon Lexical Access During Sentence Comprehension."

33 **mot juste:** See the description of how the writer searches for the perfect

match between thought and word in I. Calvino, *Six Memos for the Next Millennium* (Cambridge, MA: Harvard University Press, 1988).

34 **"there are as many connections"**: D. Eagleman, *Incognito: The Secret Lives of the Brain* (New York: Viking Press, 2011), 1.

34 **as many things are happening**: Even though I must present these processes more linearly when I describe them, the reality is a set of dynamic interactions among them that we continue to learn about. See excellent descriptions in Seidenberg, *Language at the Speed of Sight* and L. Waters, "Time for Reading," *The Chronicle of Higher Education* 53, no. 23 (February 9, 2007): B6.

LETTER THREE:
DEEP READING

35 **"I think that reading"**: M. Proust, *On Reading*, ed. J. Autret, trans. W. Burford (New York: Macmillan, 1971; originally published 1906), 48.

36 **Gina Kuperberg and Phillip Holcomb**: See in particular Kuperberg's use of multiple imaging (multimodal) methods to ascertain both the time course and the spatial information about when and what structures are involved when we read words. For example, in semantic research, she and her team use fMRI to capture a neuroanatomical picture of the networks underlying the meanings of words and both MEG and ERPs (see next note) to capture the time sequence involved. See E. F. Lau, A. Gramfort, M. S. Hämäläinen, and G. R. Kuperberg, "Automatic Semantic Facilitation in Anterior Temporal Cortex Revealed Through Multimodal Neuroimaging," *The Journal of Neuroscience* 33, no. 43 (October 23, 2013): 17174–81. See also work on ERPs in reading in J. Grainger and P. J. Holcomb, "'Watching the Word Go By': On the Time-course of Component Processes in Visual Word Recognition," *Language and Linguistics Compass* 3, no. 1 (2009): 128–56.

36 **N400 response**: The neuroscientist Marta Kutas has conducted decades of work using a form of imaging called event-related brain potential (ERP) that measures electrical activity in particular regions in milliseconds of time. The N400 is a particular form of electrical activity at around 400 milliseconds that occurs in certain areas of the brain. It is best known in its occurrence when we derive the meanings of words, particularly when they surprise our predictions. Kutas describes the N400 as "an electrical snapshot of the intersection of a feedforward flow of stimulus-driven activity with the . . . dynamically active landscape that is semantic memory." See M. Kutas and K. D. Federmeier, "Thirty Years and Counting: Finding Meaning in the N400 Component of the Event-Related Brain Potential (ERP)," *Annual Review of Psychology* 62 (2011): 621–47.

36 **when we read words**: A. Clark, "Whatever Next? Predictive Brains,

Situated Agents, and the Future of Cognitive Science," *Behavioral and Brain Sciences* 36, no. 3 (June 2013): 181–214.

37 **"proactive" predictions**: See G. R. Kuperberg, "The Proactive Comprehender: What Event-Related Potentials Tell Us About the Dynamics of Reading Comprehension," in *Unraveling Reading Comprehension: Behavioral, Neurobiological, and Genetic Components*, ed. B. Miller, L. E. Cutting, and P. McCardle (Baltimore: Paul Brookes, 2013), 176–92.

38 **"Find a Bible right now"**: F. S. Collins, *The Language of God: A Scientist Presents Evidence for Belief* (New York: Free Press, 2006), 150.

38 **"Despite twenty-five centuries"**: Ibid., 153.

39 **"a quality of attention"**: W. Stafford, "For People with Problems About How to Believe," *The Hudson Review* 35, no. 3 (September 1982): 395.

40 **"at the end there is light"**: J. Steinbeck, *East of Eden* (New York: Viking Books, 1952), 269.

40 **"When we reflect that 'sentence'"**: W. Berry, *Standing by Words* (Berkeley, CA: Counterpoint, 1983), 53.

41 **what we "see"**: P. Mendelsund, *What We See When We Read* (New York: Vintage, 2014).

41 **"Open a book and a voice speaks"**: M. Robinson, "Humanism," in *The Givenness of Things* (New York: Farrar, Straus and Giroux, 2015), 15.

41 **"For sale: baby shoes, never worn"**: Although there has been some dispute, Hemingway claimed the story to be true, with this shortest story as its result.

42 **"Only connect"**: E. M. Forster, *Howard's End* (London: Edward Arnold, 1910), chap. 22.

43 **"Passing over"**: J. S. Dunne, *Eternal Consciousness* (Notre Dame, IN: University of Notre Dame Press, 2012), 39.

43 **"That 'fruitful miracle'"**: J. S. Dunne, *Love's Mind: An Essay on Contemplative Life* (Notre Dame, IN: University of Notre Dame Press, 1993).

43 **Gish Jen**: Gish Jen attempts to illumine this principle both in her novels such as *World and Town*, in which her pitch-perfect voices of "others" bring them to life, and in her newest nonfiction explorations of the East-West culture gap, in which "other" can have very different meanings in a culture. See, in particular, her novels *World and Town*, *Mona in the Promised Land*, and *Typical American*; her short-story collection *Who's Irish?*; and her nonfiction *Tiger Writing: Art, Culture, and the Interdependent Self* and most recently *The Girl at the Baggage Claim: Explaining the East-West Culture Gap*.

45 **"We read to know"**: Quoted in J. Dunne, *A Vision Quest* (Notre Dame, IN: University of Notre Dame Press, 2006), 70.

46 **"I am not ashamed"**: Niccolò Machiavelli to Francesco Vettori, letter, December 10, 1513, in *Machiavelli and His Friends: Their*

Personal Correspondence, ed. J. Atkinson and D. Sices (Dekalb: Northern Illinois University Press, 1996).

46 "looking at my fifty friends": Quoted in S. Wasserman, "Steve Wasserman on the Fate of Books After the Age of Print," Truthdig, March 5, 2010, https://www.truthdig.com/articles/steve-wasserman-on-the-fate-of-books-after-the-age-of-print/.

47 "specialist in empathy": Interview of former president Barack Obama by Marilynne Robinson. In M. Robinson, *The Givenness of Things* (New York: Farrar, Straus and Giroux, 2015), 289.

47 "as dangerous a development": Ibid.

47 "It has to do with empathy": Ibid., quoted in N. Dames, "The New Fiction of Solitude," *The Atlantic*, April 2016, 94.

48 "I finally weep": L. Berlin, "A Manual for Cleaning Women," in *A Manual for Cleaning Women: Selected Stories* (New York: Picador, 2016), 38.

48 *Christ Actually*: J. Carroll, *Christ Actually: The Son of God for the Secular Age* (New York: Penguin, 2015).

49 Dietrich Bonhoeffer: Translations of Bonhoeffer's works in English have been published by Simon and Schuster and include *Letters and Papers from Prison*; *Ethics*; *Creation and Fall/Temptation*; *The Martyred Christian*; and *The Cost of Discipleship*. The most accessible biography is E. Metaxas, *Bonhoeffer: Pastor, Martyr, Prophet, Spy* (Nashville: Thomas Nelson, 2010). The first and most comprehensive biography is the English translation of Eberhard Bethge's *Dietrich Bonhoeffer: A Biography* (Minneapolis: Fortress Press, 2000).

49 "Only he who": Quoted in Metaxas, *Bonhoeffer*, 37.

50 Sherry Turkle described a study: See S. H. Konrath, E. H. O'Brien, and C. Hsing, "Changes in Dispositional Empathy in American College Students over Time: A Meta-analysis," *Personality and Social Psychology Review* 15, no. 2 (May 2011): 180–98.

50 loss of empathy: S. Turkle, *Reclaiming Conversation: The Power of Talk in a Digital Age* (New York: Penguin, 2015), 171–72.

50 Brain-imaging research: See especially B. C. Bernhardt and T. Singer, "The Neural Basis of Empathy," *Annual Review of Neuroscience* 35 (2012): 1–23. See also work by Bruce Miller and colleagues at UCSF.

51 some mirror neurons: See the work by Leonardo Fogassi and his colleagues, e.g., E. Kohler, C. Keysers, M. A. Umiltà, et al., "Hearing Sounds, Understanding Actions: Action Representation in Mirror Neurons," *Science* 297, no. 5582 (August 2002): 846–48; P. F. Ferrari, V. Gallese, G. Rizzolatti, and L. Fogassi, "Mirror Neurons Responding to the Observation of Ingestive and Communicative Mouth Actions in the Monkey Ventral Premotor Cortex," *European Journal of Neuroscience* 17, no. 8 (April 2003): 1703–14.

51 "Your Brain on Jane Austen": N. Phillips, "Neuroscience and the Literary History of Mind: An Interdisciplinary Approach to Attention in Jane Austen," lecture, Carnegie Mellon University, March 4, 2013.

52 **networks in the areas/Emma Bovary's silken skirt:** See the fascinating work by S. Lacey, R. Stilla, and K. Sathian, "Metaphorically
Feeling: Comprehending Textural Metaphors Activates Somatosensory Cortex," *Brain and Language* 120, no. 3 (March 2012):
416–21. See also F. Pulvermüller, "Brain Mechanisms Linking Language and Action," *Nature Reviews Neuroscience* 6, no. 7 (July
2005): 576–82; H. M. Chow, R. A. Mar, Y. Xu, et al., "Embodied
Comprehension of Stories: Interactions Between Language Regions
and Modality-Specific Mechanisms," *Journal of Cognitive Neuroscience* 26, no. 2 (February 2014): 279–95.

52 **the process of taking on:** K. Oatley, "Fiction: Simulation of Social
Worlds," *Trends in Cognitive Sciences* 20, no. 8 (August 2016):
618–28.

52 **"moral laboratory":** F. Hakemulder, *The Moral Laboratory: Ex-
periments Examining the Effects of Reading Literature on Social
Perception and Moral Self-Concept* (Amsterdam, Netherlands:
John Benjamins Publishing Company, 2000).

53 **"My guess is that":** J. Smiley, *13 Ways of Looking at the Novel*
(New York: Knopf, 2005), 177.

54 **"Who is each one of us":** I. Calvino, *Six Memos for the Next Millennium* (Cambridge, MA: Harvard University Press, 1988), 124.

54 **reading is cumulative:** A. Manguel, *A History of Reading* (New
York: Penguin, 1996).

55 **it may be possible:** See discussion in R. Kurzweil, *The Singularity Is
Near: When Humans Transcend Biology* (New York: Viking, 2005).

56 **"When the mind is braced":** R. W. Emerson, "The American Scholar,"
in Emerson, *Emerson: Essays and Lectures* (New York: Library of
America, 1983 reprint), 59.

56 **"Matthew Effect":** See K. E. Stanovich, "Matthew Effects in Reading:
Some Consequences of Individual Differences in the Acquisition of Literacy," *Reading Research Quarterly* 21, no. 4 (Fall 1986): 360–407.

57 **we run the risk:** See the discussion about the varied ways that information can be used in a digital milieu in B. Stiegler, Goldsmith
Lectures, University of London, April 14, 2013.

57 **"It would be a shame":** E. Tenner, "Searching for Dummies," *New
York Times*, March 26, 2006.

57 **"The present situation":** Remarks by G. Beasley, Conference for Libraries, Alberta, Canada, October 2014.

57 **"Chance comes only":** L. Pasteur, lecture, University of Lille, France,
December 7, 1852.

57 **STEM:** Note: this common acronym refers to science, technology,
engineering, and mathematics.

58 **"Computer, did we bring":** *Wired Staff Magazine* (November 1,
2006) asked various writers to provide their own version of the six-
word short story in the tradition of Ernest Hemingway. The science
fiction writer Eileen Gunn submitted this one.

58 "Without concepts there can be no thought": D. Hofstadter and E. Sander, *Surfaces and Essences: Analogy as the Fuel and Fire of Thinking* (New York: Basic Books, 2013), 3.

59 Leonardo Fogassi from Parma: See, e.g., Kohler et al., "Hearing Sounds, Understanding Actions"; also Ferrari et al., "Mirror Neurons Responding to the Observation of Ingestive and Communicative Mouth Actions in the Monkey Ventral Premotor Cortex."

59 Mark Greif: See M. Greif, *Against Everything: Essays* (New York: Pantheon, 2016). Do not be fooled by the title; Greif wants us to examine why we do what we do so that we know what our lives are "for."

59 "one wild and precious life": From Mary Oliver's poem "A Summer Day," Poem 133 in *Poetry 180: A Poem a Day for American High Schools*, hosted by Billy Collins, U.S. Poet Laureate, 2001–2003, http://www.loc.gov/poetry/180/133.html.

61 Widely distributed networks: L. Aziz-Zadeh, J. T. Kaplan, and M. Iacoboni, "'Aha!': The Neural Correlates of Verbal Insight Solutions," *Human Brain Mapping* 3, no. 30 (March 2009): 908–16.

61 one self-generated hypothesis after another: Ibid.

63 "What exactly is critical thinking?": M. Edmundson, *Why Read?* (New York: Bloomsbury, 2004), 43.

63 Halbertal's work: In that moment I suddenly understood Professor Halbertal's remarkable combination of incisive thought and personal gentleness more clearly: it embodies the use of long-honed knowledge in an immediate, critical analysis of new information; a deep respect for other positions; and the expectation of one's own personal conclusions. See in particular M. Halbertal, *People of the Book: Canon, Meaning, and Authority* (Cambridge, MA: Harvard University Press, 1997); M. Halbertal, *Maimonides: Life and Thought* (Princeton, NJ: Princeton University Press, 2014).

64 young boys rocking, praying, singing: I know of no better work that describes the extraordinary dimensions involved in singing the Torah than the work by Rabbi Jeffrey Summit, *Singing God's Words: The Performance of Biblical Chant in Contemporary Judaism* (Oxford, UK: Oxford University Press, 2016). The passages "Impact of a Master Reader" and "What are You Doing During the Torah Reading?," 202–06, have been particularly inspiring to me.

64 No one interpretation was assumed: I am also indebted to Barry Zuckerman for his discussions about how the very minimalism of lines from the Torah creates a basis for rich interpretation.

64 "An insight is a fleeting glimpse": J. Lehrer, "The Eureka Hunt," *The New Yorker*, July 28, 2008.

65 "Realization comes in a flash": M. P. Lynch, *The Internet of Us: Knowing More and Understanding Less in the Age of Big Data* (New York: Liveright, 2016), 177.

65 *Stories of God*: R. M. Rilke, *Stories of God*, trans. M. D. H. Norton (New York: W. W. Norton & Company, 1963).

66 Marilynne Robinson's *Gilead*: M. Robinson, *Gilead* (New York: Farrar, Straus and Giroux, 2007).

66 "It might be stated": A. Dietrich and R. Kanso, "A Review of EEG, ERP, and Neuroimaging Studies of Creativity and Insight," *Psychological Bulletin* 136, no. 5 (September 2010): 822–48.

67 "holding-ground for the contemplation": P. Davis, *Reading and the Reader: The Literary Agenda* (Oxford, UK: Oxford University Press, 2013), 8–9.

67 "neuronal workspace": S. Dehaene, *Reading in the Brain: The New Science of How We Read* (New York: Viking, 2009), 9.

67 "that invisible generative place": From William James. P. Davis, *Reading and the Reader*, 293.

67 "quarry" of language and thought: See the use of *quarry* in Emerson, "The American Scholar," 56.

LETTER FOUR:
"WHAT WILL BECOME OF THE READERS WE HAVE BEEN?"

69 "What will become": V. Klinkenborg, "Some Thoughts About E-Reading," *New York Times*, April 14, 2010.

69 "In common things": W. Wordsworth, "A Poet's Epitaph," Wikisource, from *Lyrical Ballads*, vol 2.

69 "As the devotion of a life": J. S. Dunne, *Love's Mind: An Essay on Contemplative Life* (Notre Dame, IN: University of Notre Dame Press, 1993), 3.

69 Sylvia Judson: S. S. Judson, *The Quiet Eye: A Way of Looking at Pictures* (Washington, DC: Regnery, 1982).

70 "quality of attention": W. Stafford, "For People with Problems About How to Believe," *The Hudson Review* 35, no. 3 (September 1982): 395.

70 Judith Shulevitz suggests: J. Shulevitz, *The Sabbath World: Glimpses of a Different Order of Time* (New York: Random House, 2010).

70 Frank Schirrmacher: Frank Schirrmacher, personal correspondence, August 2009.

70 *novelty bias*: A term used by Daniel Levitin in *The Organized Mind: Thinking Straight in the Age of Information Overload* (New York: Dutton, 2014).

71 A recent study by Time Inc.: See the discussion of various studies including that by Common Sense Media in N. Baron, *Words Onscreen: The Fate of Reading in a Digital World* (Oxford, UK: Oxford University Press, 2014), in particular 143–44.

71 Katherine Hayles characterized: N. K. Hayles, "Hyper and Deep Attention: The Generational Divide in Cognitive Modes," *Profession* 13 (2007): 187–99.

71 low-level threshold of boredom: See the discussion in C. Steiner-Adair, *The Big Disconnect: Protecting Childhood and Family Rela-*

tionships in the Digital Age (New York: HarperCollins, 2013). See also Baron, *Words Onscreen*, 221.

71 **continuous partial attention**: See L. Stone, "Beyond Simple Multitasking: Continuous Partial Attention," November 30, 2009, https://lindastone.net/2009/11/30/beyond-simple-multi-tasking-continuous-partial-attention/.

72 **attentional "deficits" pertain to us all**: See the discussion of Hallowell in Steiner-Adair, *The Big Disconnect*.

72 **we inhabit a world of distraction**: See the treatment of these issues in D. L. Ulin, *The Lost Art of Reading: Why Books Matter in a Distracted Time* (Seattle, WA: Sasquatch Books, 2010). See also M. Jackson, *Distracted: The Erosion of Attention and the Coming Dark Age* (Amherst, NY: Prometheus Books, 2008).

72 **Global Information Industry Center**: See the discussion of the study and the quote by R. Bohn in Ulin, *The Lost Art of Reading*, 81.

72 **"I think one thing is clear"**: Ibid.

73 **A few years later**: The director and esteemed poet Dana Gioia commissioned several reports with differing results; see, e.g., *Reading at Risk*, 2004, and *Reading on the Rise*, 2008. As of 2012, the NEA figures indicated that 58 percent of US adults had engaged in some form of literary activity, such as reading a book, over the previous year.

73 **worry that the novel**: See J. Smiley, *13 Ways of Looking at the Novel* (New York: Knopf, 2005), 177.

74 **"pursue a present"**: W. Benjamin, *Illuminations: Essays and Reflections* (New York: Schocken Books, 1968). Quoted in J. Dunne, *Love's Mind: An Essay on Contemplative Life*, 14.

74 **"a distraction, a diversion"**: Quoted in Ulin, *The Lost Art of Reading*, 62.

74 **"Swimming in entertainment"**: M. Edmundson, *Why Read?* (New York: Bloomsbury, 2004), 16.

75 **Recall the imaging study**: N. Phillips, "Neuroscience and the Literary History of Mind: An Interdisciplinary Approach to Attention in Jane Austen," lecture, Carnegie Mellon University, March 4, 2013.

75 **"masquerading as being in the know"**: Ulin, *The Lost Art of Reading*, 34.

76 **"To be a moral human being"**: M. Popova, "Susan Sontag on Storytelling, What It Means to Be a Moral Human Being, and Her Advice to Writers," *Brain Pickings*, March 30, 2015.

76 **"skimming" is the new normal**: Z. Liu, "Reading Behavior in the Digital Environment: Changes in Reading Behavior over the Past Ten Years," *Journal of Documentation* 61, no. 6 (2005): 700–12. Z. Liu, "Digital Reading," *Chinese Journal of Library and Information Science 5*, no. 1 (2012): 85–94.

77 **Naomi Baron's excellent meta-analysis**: Baron, *Words Onscreen*, 201.

77 **readers' grasp of the sequence**: See the review of these findings in

M. Wolf, *Tales of Literacy for the 21st Century* (Oxford, UK: Oxford University Press, 2016).

77 **Anne Mangen:** See overview in the following: A. Mangen and A. van der Weel, "Why Don't We Read Hypertext Novels?," *Convergence: The International Journal of Research into New Media Technologies* 23, no. 2 (May 2015): 166–81; also A. Mangen and A. van der Weel, "The Evolution of Reading in the Age of Digitisation: An Integrative Framework for Reading Research," *Literacy* 50, no. 3 (September 2016): 116–24.

78 **no significant medium-based differences:** See the varied results in literature that leave the question open, e.g., J. E. Moyer, "'Teens Today Don't Read Books Anymore': A Study of Differences in Comprehension and Interest Across Formats" (PhD diss., University of Minnesota, 2011); see also S. Eden and Y. Eshet-Alkalai, "The Effect of Format on Performance: Editing Text in Print Versus Digital Formats," *British Journal of Educational Technology* 44, no. 5 (September 2013), 846–56; R. Ackerman and M. Goldsmith, "Metacognitive Regulation of Text Learning: On Screen Versus on Paper," *Journal of Experimental Psychology: Applied* 17, no. 1 (March 2011): 18–32.

78 **Andrew Piper and David Ulin argue:** Ulin, in *The Lost Art of Reading,* quotes a provocative passage from Lewis Lapham about the effects of digital culture on sequential thought: "Sequence becomes merely additive instead of causative—the images bereft of memory, speaking to their own reflections in a vocabulary better suited to the sale of a product than to the articulation of a thought" (p. 65).

78 *technology of recurrence*: See discussion in A. Piper, *Book Was There: Reading in Electronic Times* (Chicago: University of Chicago Press, 2012), 54.

78 **"Sadly, we often atomize":** J. E. Huth, "Losing Our Way in the World," *New York Times Sunday Review,* July 20, 2013.

79 **Karin Littau:** See her extensive discussion of touch in *Theories of Reading: Books, Bodies, and Bibliomania* (Cambridge, UK: Polity Press, 2006).

80 **Nicholas Carr reminds us:** N. Carr, *The Shallows: What the Internet Is Doing to Our Brains* (New York: W. W. Norton & Company, 2010).

81 **George Miller:** See the discussion of changes in working memory in Levitin, *The Organized Mind.*

81 **"4 plus or minus 1":** Ibid.

81 **five minutes may seem:** See the discussion of changes in attention span in Baron, *Words Onscreen,* 122.

82 **diminished by more than 50 percent:** Levitin, *The Organized Mind.*

82 **"For the prose writer: success consists in":** I. Calvino, *Six Memos for the Next Millennium* (Cambridge, MA: Harvard University Press, 1988), 48.

83 "When I read a manuscript": K. Temple, "Out of the Office: The Science of Print," *Notre Dame Magazine*, December 2, 2015.

84 we "accidentally" abandon: D. Brooks, "When Beauty Strikes," *New York Times*, January 15, 2016.

84 "father forth": From the beautiful line in Gerard Manley Hopkins's poem "Pied Beauty," "He fathers forth whose beauty is past change: Praise Him." In Hopkins, *Poems and Prose of Gerard Manley Hopkins* (Baltimore: Penguin, 1933), 31.

84 "strategy of emphasis": M. Robinson, *The Givenness of Things: Essays* (New York: Farrar, Straus and Giroux, 2015), 111.

85 "In an age when other": Calvino, *Six Memos for the Next Millennium*, 45.

86 cerebrodiversity: This term, sometimes called *neurodiversity*, was used by the neuroscientist Gordon Sherman to describe how in evolution a species needs different organizations of the brain to survive. Thus in the study of dyslexia, it is important to note that this different organization of the brain *preceded* the invention of reading and was maintained genetically because of the particular skills advantaged by the dyslexic brain. See my more elaborated discussion of these issues in chaps. 7 and 8 in *Proust and the Squid: The Story and Science of the Reading Brain* (New York: HarperCollins, 2007).

86 Toni Morrison: T. Morrison, *The Origin of Others* (Cambridge, MA: Harvard University Press, 2017).

86 Aurelio Maria Mottola: Dr. Mottola directs the Vita e Pensiero publishing house in Milan, Italy, which publishes and translates some of the most important works in the humanities and social sciences for the Italian readership.

88 School Sisters of Notre Dame: This order of nuns has special significance to many neuroscientists and to those working in global education. In neuroscience research, the elderly nuns of the SSND contributed to a large research project on Alzheimer's disease and its progression, by providing both their journals that they had written over time and their brains for postmortem study. The quality of the writing in the journals yielded important clues about the onset of Alzheimer's. In addition, the SSND has been involved as teachers for many years in some of the most challenging educational environments in Africa, particularly in Liberia. See an amazing account in Sr. Mary Leonora Tucker, *I Hold Your Foot: The Story of My Enduring Bond with Liberia* (Lulu Publishing Services, 2015). See also the last chapter in Wolf, *Tales of Literacy*.

88 "Everything proceeds in geometric progression": A. Manguel, *A History of Reading* (New York: Penguin, 1996).

89 During my graduate courses: As a graduate student, I studied linguistics, particularly the development of language, with Carol Chomsky at Harvard University and participated in seminars on language and political thought with Noam Chomsky and his colleagues at MIT.

89 **written language not only reflects**: L. Vygotsky, *Thought and Language* (Cambridge, MA: MIT Press, 1986).

90 **"How was it"**: G. Eliot, *Middlemarch* (New York: Penguin Classics, 1998), 51.

91 *readability formulae*: Jeanne Chall conducted some of the most important reading research in the twentieth century; see particularly her books *Learning to Read: The Great Debate* (New York: McGraw-Hill, 1967), which analyzed the largest corpus of available data of different reading methods and concluded that code-based or phonics methods were better for most children, and *Stages of Reading Development* (New York: McGraw-Hill, 1983). Her earlier work on readability formulae was conducted to help children receive the most age-appropriate materials to read.

93 **"grit"**: A. Duckworth, *Grit: The Power of Passion and Perseverance* (New York: Simon and Schuster, 2016).

94 **"As one master"**: J. Howard, "Internet of Stings," *Times Literary Supplement*, November 30, 2016, 4.

95 **"If men learn this"**: Plato, *Phaedrus* (Princeton, NJ: Princeton University Press, 1961), 274.

95 **our intellectual evolution**: W. Ong, *Orality and Literacy: The Technologizing of the Word*, 2nd ed. (New York: Routledge, 2002).

97 **It remains for me a pitiful tale**: I might never have decided to describe this story had it not been for two probing interviews: one with Michael Rosenwald of the *Washington Post* (see his article "Serious Reading Takes a Hit from Online Scanning and Skimming, Researchers Say," April 6, 2014), the other with Maria Konnikova at *The New Yorker* (see "Being a Better Online Reader," July 16, 2014). Rosenwald wrote that his story elicited so much reaction from readers that the *Post* decided to analyze exactly how many online readers had finished it: approximately 30 percent!

97 **"acquired the status"**: Calvino, *Six Memos for the Next Millennium*, 37.

99 *Magister Ludi*: Hesse wrote *Magister Ludi*, or the *Glass Bead Game*, in German (*Das Glasperlenspiel*) over the course of many years. It was rejected for publication in Germany because of its anti-Fascist views and finally published in Switzerland in 1943. Set in a post-apocalyptic twenty-third century, the novel follows the life of Josef Knecht, who becomes an elite secular monk in an order committed to preserving the knowledge of the major disciplines through an extraordinarily complex game: the glass bead game.

100 **I was Ionesco's rhinoceros, too**: One of the more haunting plays in the Theater of the Absurd, *Rhinoceros* by Eugène Ionesco (1959), depicts how a group of people change their view of a rhinoceros from grotesque to beautiful, the more rhinoceri there are and the more they come to dominate their lives. It is a cautionary tale like few others about how human beings are influenced.

101 "the former had more velocity": A. Fadiman, ed., *Rereadings: Seventeen Writers Revisit Books They Love* (New York: Farrar, Straus and Giroux, 2005).

101 Naomi Baron predicted: Baron, *Words Onscreen*.

102 In Ong's terms: Ong, *Orality and Literacy*.

103 "Like pleated fabric": A. Goodman, "Pemberley Previsited," in *Rereadings*, A. Fadiman, ed., 164.

103 "go very far": Federico García Lorca, *The Selected Poems of Federico García Lorca* (New York: New Directions, 1955), quoted in Dunne, *Love's Mind*, 82.

104 "Among the many worlds": H. Hesse, *My Belief: Essays on Life and Art* (New York: Farrar, Straus and Giroux, 1974).

LETTER FIVE:
THE RAISING OF CHILDREN IN A DIGITAL AGE

105 *"Children are a sign"*: Pope Francis, homily, Manger Square, Bethlehem, May 25, 2014, https://w2.vatican.va/content/francesco/en/homilies/2014/documents/papa-francesco_20140525_terra-santa-omelia-bethlehem.html.

105 "Every medium has its strengths": P. M. Greenfield, "Technology and Informal Education: What Is Taught, What Is Learned," *Science* 323, no. 5910 (Jan. 2, 2009): 71.

106 "Mine own, and not mine own": W. Shakespeare, *A Midsummer Night's Dream*.

107 the more exposure to: "Repeated use of a particular media form will help to internalize the medium-specific representational skills that it uses." See K. Subrahmanyam, M. Michikyan, C. Clemmons, et al., "Learning from Paper, Learning from Screens: Impact of Screen Reading and Multitasking Conditions on Reading and Writing Among College Students," *International Journal of Cyber Behavior, Psychology and Learning* 3, no. 4 (October–December 2013): 1–27.

108 In a 2015 RAND report: See L. Guernsey and M. H. Levine, *Tap, Click, Read: Growing Readers in a World of Screens* (San Francisco: Jossey-Bass, 2015), 184.

109 "grasshopper mind": M. Weigel and H. Gardner, "The Best of Both Literacies," *Educational Leadership* 66, no. 6 (March 2009): 38–41.

109 "hop from point to point": Ibid.

109 "Humans will work": D. Levitin, *The Organized Mind: Thinking Straight in the Age of Information Overload* (New York: Dutton, 2014), 170.

110 "multitasking creates": Ibid., 96.

110 the most commonly heard complaint: C. Steiner-Adair, *The Big Disconnect: Protecting Childhood and Family Relationships in the Digital Age* (New York: HarperCollins, 2013).

110 **"dream bird that hatches"**: Quoted in J. S. Dunne, *Love's Mind: An Essay on Contemplative Life* (Notre Dame, IN: University of Notre Dame Press, 1993), 16.

111 **"If they become addicted"**: Steiner-Adair, *The Big Disconnect*, 54.

112 **the child's motor cortex enhances**: L. Fogassi, panel discussion, The Reading Brain in a Digital Culture, Spoleto, Italy, July 7, 2016.

112 **"Talk of addiction is not hyperbole"**: Steiner-Adair, *The Big Disconnect*, 6.

112 *The Chalk Artist*: A. Goodman, *The Chalk Artist* (New York: Dial Press, 2017).

113 **generation of distracted children**: Andrew Piper said something similar in *Book Was There: Reading in Electronic Times* (Chicago: University of Chicago Press, 2012), 46.

113 **Russell Poldrack and his team**: Poldrack has conducted multiple, highly influential articles on the negative effects of multitasking. See, e.g., K. Foerde, B. J. Knowlton, and R. A. Poldrack, "Modulation of Competing Memory Systems by Distraction," *PNAS* 103, no. 31 (Aug. 1, 2006): 11778–83. Newer work, however, shows some important differences for digitally raised youth who are trained in particular tasks. See K. Jimura, F. Cazalis, E. R. Stover, and R. A. Poldrack, "The Neural Basis of Task Switching Changes with Skill Acquisition," *Frontiers in Human Neuroscience* 8 (May 22, 2014): 339, 1–9.

113 **inability of most humans to switch**: Jimura et al., "Neural Basis of Task Switching Changes with Skill Acquisition," 1–9.

115 **"Our working memory"**: M. Jackson, *Distracted: The Erosion of Attention and the Coming Dark Age* (Amherst, NY: Prometheus Books, 2008), 90.

115 **Maria de Jong and Adriana Bus**: See M. T. de Jong and A. G. Bus, "Quality of Book-Reading Matters for Emergent Readers: An Experiment with the Same Book in a Regular or Electronic Format," *Journal of Educational Psychology* 94, no. 1 (2002): 145–55.

116 **Joan Ganz Cooney Center and the MacArthur Foundation**: See in particular Guernsey and Levine, *Tap, Click, Read*; L. M. Takeuchi and S. Vaala, *Level Up Learning: A National Survey on Teaching with Digital Games* (New York: Joan Ganz Cooney Center at Sesame Workshop, 2014). See also MacArthur Foundation Reports on Digital Media and Learning: e.g., J. P. Gee, *New Digital Media and Learning as an Emerging Area and "Worked Examples" as One Way Forward* (Cambridge, MA: MIT Press, 2009); M. Ito, H. A. Horst, M. Bittanti, et al., *Living and Learning with New Media: Summary of Findings from the Digital Youth Project* (Cambridge, MA: MIT Press, 2009); C. James, *Young People, Ethics, and the New Digital Media: A Synthesis from the GoodPlay Project* (Cambridge, MA: MIT Press, 2009); J. Kahne, E. Middaugh, and C. Evans, *The Civic Potential of Video Games* (Cambridge, MA: MIT Press, 2009).

116 **Kathy Hirsh-Pasek and Roberta Golinkoff:** J. Parish-Morris, N. Mahajan, K. Hirsh-Pasek, et al., "Once upon a Time: Parent-Child Dialogue and Storybook Reading in the Electronic Era," *Mind, Brain, and Education* 7, no. 3 (September 2013): 200–11. K. McNab and R. Fielding-Barnsley, "Digital Texts, iPads, and Families: An Examination of Families' Shared Reading Behaviours," *International Journal of Learning: Annual Review* 20 (2013), 53–62; Takeuchi and Vaala, *Level Up Learning*; Guernsey and Levine, *Tap, Click, Read*, 18.

116 **"The highly enhanced e-book":** Guernsey and Levine, *Tap, Click, Read*, 184.

119 **She argues that our cultural inventions:** See A. Winter, *Memory: Fragments of a Modern History* (Chicago: University of Chicago Press, 2012); also see my review of the book: M. Wolf, "Memory's Wraith," *The American Interest* 9, no. 1 (Aug. 11, 2013): 85–89.

119 **"While narratives are":** S. Greenfield, *Mind Change: How Digital Technologies Are Leaving Their Mark on Our Brains* (New York: Random House, 2015), 243.

120 **"arrive in the brain":** Ibid., 46–47.

121 **when there is too much information overload:** See discussion in Jackson, *Distracted*, esp. 79–80.

121 **Katherine Hayles sharpens:** N. K. Hayles, "Hyper and Deep Attention: The Generational Divide in Cognitive Modes," *Profession* 13 (2007): 187–99.

122 **"habituating us to ever faster":** E. Hoffman, *Time* (New York: Picador, 2009), 12.

123 **"I worry that the level":** Quoted in Greenfield, *Mind Change*, 26.

124 **"Imagine in the future":** Ibid., 206.

124 **Ray Kurzweil:** R. Kurzweil, *The Singularity Is Near: When Humans Transcend Biology* (New York: Viking, 2005). See particularly the discussion at 4, 128.

125 **Tristan Harris:** "Your Phone Is Trying to Control Your Life," interview with Tristan Harris, *PBS NewsHour*, January 30, 2017. See also B. Bosker, "The Binge Breaker," *The Atlantic*, November 2016.

125 **Josh Elman:** Bosker, "The Binge Breaker."

125 **"Never before in history":** See discussion in Bosker, "The Binge Breaker."

125 **would agree with this responsibility:** As discussed in Bosker, "The Binge Breaker," Larry Page, the CEO of Google, discussed Harris's concepts of how Google could better address those criticisms. Later Harris worked specifically on the incorporation of "ethical design" at Google, before leaving to found his Time Well Spent initiative. In 2015, Google changed its guiding principle to "Do the Right Thing," NBC News, Tech News, Jan. 19, 2018.

126 **You and I can hold:** The original quote comes from "The Crack-Up" (1936): "The test of a first-rate intelligence is the ability to hold

two opposed ideas in the mind at the same time, and still retain the ability to function." F. Scott Fitzgerald, "The Crack-Up," *Esquire*, March 7, 2017, http://www.esquire.com/lifestyle/a4310/the-crack-up/.

126 **work being conducted**: See Guernsey and Levine, *Tap, Click, Read*, and Baron, *Words Onscreen*. For work by the European based E-READ network, see M. Barzillai, J. Thomson, and A. Mangen, "The Influence of E-books on Language and Literacy Development," in *Education and New Techologies: Perils and Promises for Learners*, ed. K. Sheehy and A. Holliman (London: Routledge, forthcoming).

127 **"build the habits of mind"**: Guernsey and Levine, *Tap, Click, Read*, 40.

LETTER SIX:
FROM LAPS TO LAPTOPS IN THE FIRST FIVE YEARS

128 **"Is the real barrier"**: L. Guernsey and M. H. Levine, *Tap, Click, Read: Growing Readers in a World of Screens* (San Francisco: Jossey-Bass, 2015), 8–9.

128 **"Books and screens"**: A. Piper, *Book Was There: Reading in Electronic Times* (Chicago: University of Chicago Press, 2012), ix.

128 **"room where it happens"**: A well-deserved nod to the Broadway musical *Hamilton*.

128 **"crook of an arm"**: M. Wolf, *Proust and the Squid: The Story and Science of the Reading Brain* (New York: HarperCollins, 2007), 81. See chap. 4 for a much more expansive discussion.

129 **comfortable adaptation of fMRI**: See S. Dehaene, *Consciousness and the Brain: Deciphering How the Brain Decodes Our Thoughts* (New York: Penguin, 2009).

130 **"The crucial condition"**: C. Taylor, *The Language Animal: The Full Shape of the Human Linguistic Capacity* (Cambridge, MA: Harvard University Press, 2016), 177.

130 **New brain-imaging research**: See J. Hutton, "Stories and Synapses: Home Reading Environment and Brain Function Supporting Emergent Literacy," presentation to the Reach Out and Read Conference, Boston, May 2016. See also T. Horowitz-Kraus, R. Schmitz, J. S. Hutton, and J. Schumacher, "How to Create a Successful Reader? Milestones in Reading Development from Birth to Adolescence," *Acta Paediatrica* 106, no. 4 (April 2017).

131 **"All shall be well"**: See Reverend Mother Julia Gatta's moving descriptions of Julian of Norwich in *The Pastoral Art of the English Mystics* (originally published as *Three Spiritual Directors for Our Time* [Cambridge, MA: Cowley Publishers, 1987]).

133 **"The digital page"**: A. Piper, *Book Was There: Reading in Electronic Times* (Chicago: University of Chicago Press, 2012), 54.

134 **Common Sense Media**: See *Children, Teens, and Reading: A Common*

Sense Media Research Brief, May 12, 2014, https://www.common
sensemedia.org/research/children-teens-and-reading. Also reported in
C. Alter, "Study: The Number of Teens Reading for Fun Keeps Declining," *Time*, May 2, 2014.

134 **dip in parent-child reading**: Despite these important initiatives and
despite the fact that over 80 percent of even slightly older children
(between six and eight years of age) wish their parents would still
read to them, there has been a decline in this simple, invaluable
contribution to the child's formation of reading and a simultaneous
increase in children's digital time. http://www.bringmeabook.org.

134 **one of the single most important**: Beginning in the 1970s with landmark studies by Carol Chomsky and Charles Read (see discussion
in Wolf, *Proust and the Squid*) and continuing to the present with
work by Catherine Snow and her colleagues, this simple intervention by parents has remained one of the best predictors of how well
children will read later in life.

135 **Reach Out and Read**: See http://www.reachoutandread.org.

135 **Born to Read**: See http://www.borntoread.org.

135 **Bring Me a Book**: See http://www.bringmeabook.org.

135 **physical books—not apps or e-books**: There is a burgeoning re-
search base on this topic for parents. See N. Kucirkova and B.
Zuckerman, "A Guiding Framework for Considering Touchscreens
in Children Under Two," *International Journal of Child-Computer
Interaction* 12, issue C (April 2017): 46–49; N. Kucirkova and K.
Littleton, *The Digital Reading Habits of Children* (London: Book
Trust, 2016); J. S. Radesky, C. Kistin, S. Eisenberg, et al., "Parent
Perspectives on Their Mobile Technology Use: The Excitement and
Exhaustion of Parenting While Connected," *Journal of Develop-
mental & Behavioral Pediatrics* 37, no. 9 (November–December
2016): 694–701; J. S. Radesky, J. Schumacher, and B. Zuckerman,
"Mobile and Interactive Media Use by Young Children: The Good,
the Bad, and the Unknown," *Pediatrics* 135, no. 1 (January 2015):
1–3; C. Lerner and R. Barr, "Screen Sense: Setting the Record
Straight: Research-Based Guidelines for Screen Use for Children
Under 3 Years Old," *Zero to Three*, May 2, 2014, https://www
.zerotothree.org/resources/1200-screen-sense-full-white-paper. See
also an earlier study: R. Needlman, L. E. Fried, D. S. Morley, et
al., "Clinic-Based Intervention to Promote Literacy: A Pilot Study,"
The American Journal of Diseases of Children 145, no. 8 (August
1991): 881–84.

135 **limited contact with digital devices**: See the corpus of work by
Kathy Hirsh-Pasek and Roberta Golinkoff, e.g., R. M. Golinkoff,
K. Hirsh-Pasek, and D. Eyer, *Einstein Never Used Flash Cards:
How Our Children Really Learn—and Why They Need to Play
More and Memorize Less* (Emmaus, PA: Rodale Books, 2003), and
new work cited in the last letter.

136 **"God made Man"**: E. Wiesel, *The Gates of the Forest* (New York: Schocken, 1996), Preface.

137 **the radius of children's activity**: S. Greenfield, *Mind Change: How Digital Technologies Are Leaving Their Mark on Our Brains* (New York: Random House, 2015), 19.

138 **"practice reacting"**: J. Gottschall, *The Storytelling Animal: How Stories Make Us Human* (Boston: Houghton Mifflin Harcourt, 2012), 67.

138 **Such a thought**: S. Pinker, *How The Mind Works* (New York: W. W. Norton & Company, 1997).

138 **"compassionate imagination"**: M. C. Nussbaum, *Cultivating Humanity: A Classical Defense of Reform in Liberal Education* (Cambridge, MA: Harvard University Press, 1997), 92.

138 **Here begins the moral laboratory**: F. Hakemulder, *The Moral Laboratory: Experiments Examining the Effects of Reading Literature on Social Perception and Moral Self-Concept* (Amsterdam, Netherlands: John Benjamins Publishing Company, 2000).

139 **Jean Berko Gleason**: Known best for her whimsical ways of eliciting morphological knowledge in children through the "Wug" test, Gleason was one of the seminal influences on twentieth-century developmental psycholinguistics. See the ninth edition of her book *The Development of Language* (New York: Pearson, 2016) with coeditor Nan Ratner Bernstein, which brings together research in this area over two decades.

140 **older research by British experts**: An older direction of research demonstrated that the rhymes of Mother Goose are one of the best preparations for focusing a child's attention on the phonemes of words. See L. Bradley and P. E. Bryant, "Categorizing Sounds and Learning to Read—A Causal Connection," *Nature* 301 (February 3, 1983): 419–21; L. Bradley and P. Bryant, *Rhyme and Reason in Spelling* (Ann Arbor: University of Michigan Press, 1985); P. Bryant, M. MacLean, and L. Bradley, "Rhyme, Language, and Children's Reading," *Applied Psycholinguistics* 11, no. 3 (September 1990): 237–52.

140 **the rhythm in music**: Cathy Moritz, Aniruddh Patel, Ola Ozernov-Palchik, and other members of my center have been conducting studies of music's relationship to reading, particularly the relationship between rhythm in music and phoneme awareness. Moritz and our group found that daily music training in kindergarten predicts better reading achievement at the end of first grade, a finding that flies in the face of the cuts in music programs around the country. Ola Ozernov-Palchik and Ani Patel are conducting more in-depth studies of the relationship between music and reading, so as to use this knowledge as the basis of prediction and intervention. See C. Moritz, S. Yampolsky, G. Papadelis, et al., "Links Between Early Rhythm Skills, Musical Training, and Phonological Awareness," *Reading and Writing* 26, no. 5 (May 2013): 739–69.

143 **well over a million:** See a selected list in Guernsey and Levine, *Tap, Click, Read.*

143 **parents always think about:** Ibid.

144 **when parents read stories on e-books:** A. R. Lauricella, R. Barr, and S. L. Calvert, "Parent-Child Interactions During Traditional and Computer Storybook Reading for Children's Comprehension: Implications for Electronic Storybook Design," *International Journal of Child-Computer Interaction* 2, no. 1 (January 2014): 17–25; S. E. Mol and A. G. Bus, "To Read or Not to Read: A Meta-analysis of Print Exposure from Infancy to Early Adulthood," *Psychological Bulletin* 137, no. 2 (March 2011): 267–96; S. E. Mol, A. G. Bus, M. T. de Jong, and D. J. H. Smeets, "Added Value of Dialogic Parent-Child Book Readings: A Meta-analysis," *Early Education and Development* 19 (2008): 7–26; O. Segal-Drori, O. Korat, A. Shamir, and P. S. Klein, "Reading Electronic and Printed Books with and without Adult Instruction," *Reading and Writing: An Interdisciplinary Journal* 23, no. 8 (September 2010): 913–30. See also M. Barzillai, J. Thomson, and A. Mangen, "The Influence of E-books on Language and Literacy Development," in *Education and New Techologies: Perils and Promises for Learners*, ed. K. Sheehy and A. Holliman (London: Routledge, forthcoming).

145 **negative influence of interactive digital books:** A. G. Bus, Z. K. Takacs, and C. A. T. Kegel, "Affordances and Limitations of Electronic Storybooks for Young Children's Emergent Literacy," *Developmental Review* 35 (March 2015): 79–97.

145 **parents can use the interactive nature:** See a fuller description in M. Wolf, S. Gottwald, C. Breazeal, et al., "'I Hold Your Foot': Lessons from the Reading Brain for Addressing the Challenge of Global Literacy," in *Children and Sustainable Development*, ed. A. Battro, P. Léna, M. Sánchez Sorondo, and J. von Braun (Cham, Switzerland: Springer Verlag, 2017). See also A. Chang's PhD dissertation, MIT Media Lab, 2011; C. Breazeal, "TinkRBook: Shared Reading Interfaces for Storytelling," IDC, June 20, 2011.

147 **the role that the human-technology interface:** See, e.g., M. A. Hearst, "'Natural' Search User Interfaces," *Communications of the ACM* 54, no. 11 (November 2011): 60–67; M. Hearst, "Can Natural Language Processing Become Natural Language Coaching?," keynote presentation, ACL, Beijing, July 2015.

147 **Carola and Marcelo Suárez-Orozco:** These UCLA researchers have contributed an extraordinary corpus of scholarly work on immigrant children, including on the cognitive flexibility in dual-language learners; see, e.g., C. Suárez-Orozco, M. M. Abo-Zena, and A. K. Marks, eds., *Transitions: The Development of the Children of Immigrants* (New York: New York University Press, 2015). See also E. Bialystok and M. Viswanathan, "Components of Executive Control with Advantages for Bilingual Children in Two Cultures,"

Cognition 112, no. 3 (September 2009): 494–500; K. Hakuta and R. M. Diaz, "The Relationship Between Degree of Bilingualism and Cognitive Ability: A Critical Discussion and Some New Longitudial Data," *Children's Language 5* (1985): 319–44; W. E. Lambert, "Cognitive and Socio-Cultural Consequences of Bilingualism," *Canadian Modern Language Review* 34, no. 3 (February 1978): 537–47; O. O. Adesope, T. Lavin, T. Thompson, and C. Ungerleider, "A Systematic Review and Meta-analysis of the Cognitive Correlates of Bilingualism," *Review of Educational Research* 80, no. 2 (2010): 207–45.

148 **our country's immigrant children:** M.-J. A. J. Verhallen, A. G. Bus, and M. T. de Jong, "The Promise of Multimedia Stories for Kindergarten Children at Risk," *Journal of Educational Psychology* 98, no. 2 (May 2006): 410–19.

LETTER SEVEN:
THE SCIENCE AND POETRY IN LEARNING
(AND TEACHING) TO READ

150 **There is nothing:** S. Deheane, *Reading in the Brain: The New Science of How We Read* (New York: Penguin, 2009), 326.

150 **And what do we learn:** M. Dirda, *Book by Book* (New York: Henry Holt, 2007), 70.

150 **"children who learn":** W. James, quoted in M. Wolf, "'As Birds Fly': Fluency in Children's Reading" (New York: Scholastic Publishing, 2001).

151 **at four he was:** B. Collins, "On Turning Ten," in *The Art of Drowning* (Pittsburgh: University of Pittsburgh Press, 1995), 48.

151 **Every national and international index:** See the sobering results of US children when compared to children around the world in Programme for International Student Assessment (PISA), http://www.oecd.org/pisa; and Amanda Ripley's discussion of PISA comparisons in *The Smartest Kids in the World: And How They Got That Way* (New York: Simon and Schuster, 2013). See also M. Seidenberg, *Language at the Speed of Sight: How We Read, Why So Many Can't, and What Can Be Done About It* (New York: Basic Books, 2017). See also the results of the 2003 National Assessment of Adult Literacy, which found that 93 million people in the United States read at basic or below basic levels.

151 **"proficient" level:** See the similarly sobering results from NAEP at http://www.nationsreportcard.gov, where more than half of the children scored at or below basic levels at each testing; also discussed at length in Seidenberg, *Language at the Speed of Sight*. See also "Children, Teens, and Reading: A Common Sense Research Brief," May 12, 2014, https://www.commonsensemedia.org/research/children-teens-and-reading. Also reported in C. Alter, "Study: The Number of Teens Reading for Fun Keeps Declining," *Time*, May 12, 2014.

152 **the relationship between:** See C. Coletti, *Blueprint for a Literate Nation: How You Can Help* (Xlibris, 2013).

152 **"Large, undereducated swaths":** Cited in Council on Foreign Relations, *U.S. Education Reform and National Security* (New York: Council on Foreign Relations, 2012); see also Seidenberg, *Language at the Speed of Sight.*

153 **children in underprivileged families:** B. Hart and T. R. Risley, "The Early Catastrophe: The 30 Million Word Gap," *American Educator* 27, no. 1 (Spring 2003): 4–9; B. Hart and T. R. Risley, *Meaningful Differences in the Everyday Experience of Young American Children* (Baltimore: Brookes Publishing, 1995).

153 **James Heckman:** J. J. Heckman, *Giving Kids a Fair Chance (A Strategy That Works)* (Cambridge, MA: MIT Press, 2013). See also a fascinating description of Heckman's results and related research in a documentary film produced by Christine Herbes-Sommers, *The Raising of America*, 2016.

153 **more comprehensive early-childhood programs:** See J. P. Shonkoff and D. A. Phillips, eds., *From Neurons to Neighborhoods: The Science of Early Childhood Development* (Washington, DC: National Academy Press, 2000); D. Stipek, "Benefits of Preschool Are Clearly Documented," *Mercury News*, August 6, 2013; D. Stipek, "No Child Left Behind Comes to Preschool," *The Elementary School Journal* 106, no. 5 (May 2006): 455–66.

153 **rejects the term *gap*:** See discussion in L. Guernsey and M. H. Levine, *Tap, Click, Read: Growing Readers in a World of Screens* (San Francisco: Jossey-Bass, 2015), 25.

155 **of the largest reading prediction studies:** O. Ozernov-Palchik, E. S. Norton, G. Sideridis, M. Wolf, N. Gaab, J. Gabrieli, et al. (2016), "Longitudinal Stability of Pre-reading Skill Profiles of Kindergarten Children: Implications for Early Screening and Theories of Reading," *Developmental Science* 20, no. 5 (September 2017): 1–18. See also O. Ozernov-Palchik and N. Gaab, "Tackling the 'Dyslexia Paradox': Reading Brain and Behavior for Early Markers of Developmental Dyslexia," *WIREs Cognitive Science* 7, no. 2 (March–April 2016): 156–76; Z. M. Saygin, E. S. Norton, D. E. Osher, et al., "Tracking the Roots of Reading Ability: White Matter Volume and Integrity Correlate with Phonological Awareness in Prereading and Early-Reading Kindergarten Children," *The Journal of Neuroscience* 33, no. 33 (Aug. 14, 2013): 13251–58.

155 **gives children with dyslexia:** See chaps. 7 and 8 in my book *Proust and the Squid: The Story and Science of the Reading Brain* (New York: HarperCollins, 2007) for an overview of dyslexia, with emphases on the creativity often found in individuals with dyslexia and why these patterns of thinking outside the box are often the sources of later success in entrepreneurs.

156 **Other researchers at UCSF School of Medicine:** See their ongoing work at the Dyslexia Center, School of Medicine, University of California, San Francisco; e.g., Johns Hopkins pediatric neurologist Martha Denckla, personal correspondence, Fall 2015.

157 **reading developed with fewer problems:** U. Goswami, "How to Beat Dyslexia," *The Psychologist* 16, no. 9 (2003): 462–65.

157 **physiological and behavioral reasons:** See Wolf, *Proust and the Squid*, chaps. 4 and 5.

158 *The Raising of America*: This important documentary on the topic, *The Raising of America*, 2016, produced by Christine Herbes-Sommers, shows the long-term deleterious effects of early deprivation, as well as the ameliorative ones of good early care.

158 **as complex a set of knowledge bases:** See the comment in L. C. Moats, *Teaching Reading Is Rocket Science* (Washington, DC: American Federation of Teachers, 1999).

159 **so-called Reading Wars:** See J. Chall, *Learning to Read: The Great Debate* (New York: McGraw-Hill, 1967), which analyzed the largest corpus of available data of different reading methods and concluded that code-based or phonics methods were better for most children. The debate about these methods has never abated and has long been referred to as the Reading Wars.

160 **federally funded research studies:** Many substantive overviews of this research have been the subject of multiple volumes over the last decade and a half, edited by past and present National Institute of Child Health and Human Development directors of research on reading and reading disabilities Peggy McCardle and Brett Miller. Other overviews have been the result of conferences on intervention research organized by the Dyslexia Foundation. See K. Pugh and P. McCardle, eds., *How Children Learn to Read: Current Issues and New Directions in the Integration of Cognition, Neurobiology and Genetics of Reading and Dyslexia Research and Practice* (New York: Psychology Press, 2009). See also P. E. McCardle and V. E. Chhabra, eds., *The Voice of Evidence in Reading Research* (Baltimore: Brookes Publishing, 2004); B. Miller, P. McCardle, and R. Long, eds., *Teaching Reading and Writing: Improving Instruction and Student Achievement* (Baltimore: Brookes Publishing, 2014); B. Miller, L. E. Cutting, and P. McCardle, eds., *Unraveling Reading Comprehension: Behavioral, Neurobiological, and Genetic Components* (Baltimore: Brookes Publishing, 2013).

160 **common core standards:** This topic is immensely important and complex and deserves far more than a cursory footnote. See the important work by individual states on this topic, e.g., California Common Core State Standards and Connecticut Common Core State Standards.

160 **"theoretical zombies":** Seidenberg, *Language at the Speed of Sight*, 271.

162 **fluency:** See more comprehensive accounts of fluency in M. Wolf and T. Katzir-Cohen, "Reading Fluency and Its Intervention," *Scientific Studies of Reading* 5, no. 3 (2001): 211–38; T. Katzir, Y. Kim, M. Wolf, et al., "Reading Fluency: The Whole Is More than the Parts," *Annals of Dyslexia* 56, no. 1 (March 2006): 51–82.

162 **A decade of research:** R. D. Morris, M. W. Lovett, M. Wolf, et al., "Multiple-Component Remediation for Developmental Reading Disabilities: IQ, Socioeconomic Status, and Race as Factors in Remedial Outcome," *Journal of Learning Disabilities* 45, no. 2 (March–April 2012): 99–127; M. W. Lovett, J. C. Frijters, M. Wolf, et al., "Early Intervention for Children at Risk for Reading Disabilities: The Impact of Grade at Intervention and Individual Differences on Intervention Outcomes," *Journal of Educational Psychology* 109, no. 7 (October 2017): 889–914. See fuller descriptions of my group's intervention, the RAVE-O reading program, in M. Wolf, C. Ullman-Shade, and S. Gottwald, "The Emerging, Evolving Reading Brain in a Digital Culture: Implications for New Readers, Children with Reading Difficulties, and Children Without Schools," *Journal of Cognitive Education and Psychology* 11, no. 3 (2012): 230–40; M. Wolf, M. Barzillai, S. Gottwald, et al., "The RAVE-O Intervention: Connecting Neuroscience to the Classroom," *Mind, Brain, and Education* 3, no. 2 (June 2009): 84–93.

162 **fluent reading involves:** Please note the expansive work in Hebrew on fluency processes not only in reading but also in emotions by Tami Katzir and her colleagues in Haifa. Daniela Traficante and her PhD student Valentina Andolfi have done impressive work on intervention for fluent comprehension of Italian in the Eureka program, modeled after the RAVE-O program in English.

163 **"Education for world citizenship":** M. C. Nussbaum, *Cultivating Humanity: A Classical Defense of Reform in Liberal Education* (Cambridge, MA: Harvard University Press, 1997), 69, 93.

166 **A large, ongoing initiative:** See overviews in C. E. Snow, "2014 Wallace Foundation Distinguished Lecture: Rigor and Realism: Doing Educational Science in the Real World," *Educational Researcher* 44, no. 9 (December 2015): 460–66; P. Uccelli, C. D. Barr, C. L. Dobbs, et al., "Core Academic Language Skills (CALS): An Expanded Operational Construct and a Novel Instrument to Chart School-Relevant Language Proficiency in Preadolescent and Adolescent Learners," *Applied Psycholinguistics* 36, no. 5 (September 2015): 1077–1109; P. Uccelli and E. P. Galloway, "Academic Language Across Content Areas: Lessons from an Innovative Assessment and from Students' Reflections About Language," *Journal of Adolescent & Adult Literacy* 60, no. 4 (January–February 2017): 395–404; P. Uccelli, E. P. Galloway, C. D. Barr, et al., "Beyond Vocabulary: Exploring Cross-Disciplinary Academic-Language Proficiency and Its Association with Reading Comprehension," *Reading Research Quarterly* 50, no. 3 (July–September 2015): 337–56.

LETTER EIGHT:
BUILDING A BILITERATE BRAIN

168 "The depth of the challenge": L. Guernsey and M. H. Levine, *Tap, Click, Read: Growing Readers in a World of Screens* (San Francisco: Jossey-Bass, 2015), 39.

168 65 percent of the jobs: A. Ross, *The Industries of the Future* (New York: Simon and Schuster, 2016).

169 *learned ignorance*: Originally written in 1440; see Nicholas of Cusa, *On Learned Ignorance*, trans. J. Hopkins (Minneapolis: Banning, 1985).

170 By the time they reach adulthood: See Ellen Bialystok's research, particularly E. Bialystok, F. I. M. Craik, D. W. Green, and T. H. Gollan, "Bilingual Minds," *Psychological Science in the Public Interest* 10, no. 3 (December 2009): 89–129.

170 Rapid Alternating Stimulus: M. Wolf and M. B. Denckla, "RAN/ RAS: Rapid Automatized Naming and Rapid Alternating Stimulus Tests" (Austin, TX: Pro-Ed, 2005).

171 groundbreaking work: See C. Goldenberg, "Congress: Bilingualism Is Not a Handicap," *Education Week*, July 14, 2015; C. Goldenberg and R. Coleman, *Promoting Academic Achievement Among English Learners: A Guide to the Research* (Thousand Oaks, CA: Corwin, 2010); A. Y. Durgunoğlu and C. Goldenberg, eds., *Language and Literacy Development in Bilingual Settings* (New York: Guilford Press, 2011).

171 Lev Vygotsky: See L. Vygotsky, *Thought and Language* (Cambridge, MA: MIT Press, 1986).

172 "guiding this peripatetic mind": M. Weigel and H. Gardner, "The Best of Both Literacies," *Educational Leadership* 66, no. 6 (March 2009): 38–41.

173 "gnys at wrk": G. L. Bissex, *Gnys at Wrk: A Child Learns to Write and Read* (Cambridge, MA: Harvard University Press, 1985).

173 write their thoughts by hand: See, e.g., S. Graham and T. Santangelo, "A Meta-analysis of the Effectiveness of Teaching Handwriting," presentation, Handwriting in the 21st Century? An Educational Summit, Jan. 23, 2012. See also work by the neurologist William Klemm.

174 "Every child should be given": M. U. Bers and M. Resnick, *The Official ScratchJr Book: Help Your Kids Learn to Code* (San Francisco: No Starch Press, 2015), 2–3.

175 She and her team demonstrate: C. Breazeal, "Emotion and Sociable Humanoid Robots," *International Journal of Human-Computer Studies* 59, nos. 1–2 (July 2003): 119–55.

175 trying to confront the cognitive challenges: M. Barzillai, J. Thomson, and A. Mangen, "The Influence of E-books on Language and Literacy Development," in *Education and New Techologies: Perils*

and Promises for Learners, ed. K. Sheehy and A. Holliman (London: Routledge, forthcoming); M. Wolf and M. Barzillai, "The Importance of Deep Reading," *Educational Leadership* 66, no. 6 (March 2009): 32–35.

176 **Thinking Reader program:** B. Dalton and D. Rose, "Scaffolding Digital Comprehension," in *Comprehension Instruction: Research-Based Best Practices*, 2nd ed., ed. C. C. Block and S. R. Parris (New York: Guilford Press, 2008), 347–61.

176 **UDL principles:** D. H. Rose and A. Meyer, *Teaching Every Student in the Digital Age: Universal Design for Learning* (Alexandria, VA: ASCD, 2002).

176 **the program incorporates:** Members of the CAST team describe a continuum of forms of support that "provide access to the content (e.g., a struggling reader may use Text to Speech support to have the text read aloud via synthetic voice or view a multimedia definition) or additional information needed to comprehend the text (e.g., an ELL may hear a word pronounced, learn the Spanish translation for that word, and write a personal association with the word)." See A. Meyer, D. Rose, and D. Gordon, *Universal Design for Learning*, (Warefield, MA: CAST Professional Publishing), 2014.

176 **One of the consistent cautions:** See S. Lefever-Davis and C. Pearman, "Early Readers and Electronic Texts: CD-ROM Storybook Features That Influence Reading Behaviors," *The Reading Teacher* 58, no. 5 (February 2005): 446–54.

177 **MacArthur Foundation:** See the extensive reports on digital tools and activities sponsored by the MacArthur Foundation Reports on Digital Media and Learning; e.g., C. N. Davidson and D. T. Goldberg, *The Future of Learning Institutions in a Digital Age* (Cambridge, MA: MIT Press, 2009); J. P. Gee, *New Digital Media and Learning as an Emerging Area and "Worked Examples" as One Way Forward* (Cambridge, MA: MIT Press, 2009); M. Ito, H. A. Horst, M. Bittanti, et al., *Living and Learning with New Media: Summary of Findings from the Digital Youth Project* (Cambridge, MA: MIT Press, 2009); C. James, *Young People, Ethics, and the New Digital Media: A Synthesis from the GoodPlay Project* (Cambridge, MA: MIT Press, 2009); H. Jenkins, *Confronting the Challenges of Participatory Culture: Media Education for the 21st Century* (Cambridge, MA: MIT Press, 2009).

177 **"digital wisdom":** J. Coiro, "Online Reading Comprehension: Challenges and Opportunities," *Texto Livre: Linguagem e Tecnologia* 7, no. 2 (2014): 30–43.

179 **have never been given training:** Guernsey and Levine, *Tap, Click, Read*, 233.

181 **A fascinating study:** As discussed by me in *Tales of Literacy*, I am wondering whether Coiro's data are showing the emergence of two differently formed reading circuits. See J. Coiro, "Predicting Read-

ing Comprehension on the Internet: Contributions of Offline Reading Skills, Online Reading Skills, and Prior Knowledge," *Journal of Literacy Research* 43, no. 4 (2011): 352–92.

181 **For slightly older children:** See S. Vaughn, J. Wexler, A. Leroux, et al., "Effects of Intensive Reading Intervention for Eighth-Grade Students with Persistently Inadequate Response to Intervention," *Journal of Learning Disabilities* 45, no. 6 (November–December 2012): 515–25.

181 **audiobooks:** See M. Rubery, *The Untold Story of the Talking Book* (Cambridge, MA: Harvard University Press, 2016).

181 **video games:** See J. Gee, *What Video Games Have to Teach Us About Learning and Literacy* (New York: Palgrave Macmillan, 2003). See also the extensive reports on digital games and activities sponsored by the MacArthur Foundation Reports on Digital Media and Learning: e.g., Gee, *New Digital Media and Learning as an Emerging Area and "Worked Examples" as One Way Forward*; Ito et al., *Living and Learning with New Media*; C. James, *Young People, Ethics, and the New Digital Media*; J. Kahne, E. Middaugh, and C. Evans, *The Civic Potential of Video Games* (Cambridge, MA: MIT Press, 2009).

182 **Meta-analyses of studies:** A. C. K. Cheung and R. E. Slavin, "The Effectiveness of Education Technology for Enhancing Reading Achievement: A Meta-analysis," Center for Research and Reform in Education, Johns Hopkins University, May 2011; A. C. K. Cheung and R. E. Slavin, "How Features of Educational Technology Applications Affect Student Reading Outcomes: A Meta-analysis," *Educational Research Review* 7, no. 3 (December 2012): 198–215; A. C. K. Cheung and R. E. Slavin, "The Effectiveness of Educational Technology Applications for Enhancing Mathematics Achievement in K–12 Classrooms: A Meta-analysis," *Educational Research Review* 9 (June 2013): 88–113; Y-C. Lan, Y-L. Lo, and Y-S. Hsu, "The Effects of Meta-cognitive Instruction on Students' Reading Comprehension in Computerized Reading Contexts: A Quantitative Meta-analysis," *Journal of Educational Technology & Society* 17, no. 4 (October 2014): 186–202; Q. Li and X. Ma, "A Meta-analysis of the Effects of Computer Technology on School Students' Mathematics Learning," *Educational Psychology Review* 22, no. 3 (September 2010): 215–43.

183 **"use of the computer may have widened":** See S. White, Y. Y. Kim, J. Chen, and F. Liu, "Performance of Fourth-Grade Students in the 2012 NAEP Computer-Based Writing Pilot Assessment: Scores, Test Length, and Editing Tools," working paper, Institute of Education Sciences, Washington, DC, October 2015.

183 **Children who have less exposure:** See discussion by Stephanie Gottwald and myself about the nonliterate child in chap. 3 of *Tales of Literacy for the 21st Century* (Oxford, UK: Oxford University Press, 2016).

183 **"Opportunity for All?":** V. Rideout and V. S. Katz, "Opportunity for All?: Technology and Learning in Lower-Income Families," Joan Ganz Cooney Center at Sesame Workshop, New York, 2016.

183 **two different kinds of digital gaps:** Ibid.; H. Jenkins, *Confronting the Challenges of Participatory Culture: Media Education for the 21st Century* (Cambridge, MA: MIT Press, 2009).

184 **"access is no longer":** Rideout and Katz, "Opportunity for All?," 7.

184 **one of the most discouraging:** Cited in Guernsey and Levine, *Tap, Click, Read.*

185 **Curious Learning:** See M. Wolf et al., "The Reading Brain, Global Literacy, and the Eradication of Poverty," *Proceedings of Bread and Brain, Education and Poverty* (Vatican City: Pontifical Academy of Social Sciences, 2014); M. Wolf et al., "Global Literacy and Socially Excluded Peoples," *Proceedings of the Emergency of the Socially Excluded* (Vatican City: Pontifical Academy of Social Sciences, 2013).

186 **Adult Literacy XPRIZE:** https://adultliteracy.xprize.org.

186 **children with the flexible:** C. Suárez-Orozco, M. M. Abo-Zena, and A. K. Marks, eds., *Transitions: The Development of the Children of Immigrants* (New York: New York University Press, 2015). See the extensive work cited in Letter Seven.

186 **"Our only world":** W. Berry, *Our Only World: Ten Essays* (Berkeley, CA: Counterpoint, 2015).

187 **"The future—any future":** P. A. McKillip, *The Moon and the Face* (New York: Berkley, 1985), 88.

187 **"I can, with one eye squinted":** B. Gooch, *Flannery: A Life of Flannery O'Connor* (New York: Little, Brown and Company, 2009), 229.

LETTER NINE:
READER, COME HOME

188 **"To read, we need":** D. L. Ulin, *The Lost Art of Reading: Why Books Matter in a Distracted Time* (Seattle, WA: Sasquatch Books, 2010), 34, 16, 150.

188 **"Past a certain scale":** W. Berry, *Standing by Words: Essays* (Washington, DC: Shoemaker & Hoard, 2005), 60–61.

189 **a good society has three lives:** Aristotle, *The Nicomachean Ethics*, trans. H. Rackham (New York: William Heinemann, 1926).

189 **understanding of *leisure*:** J. Pieper, *Leisure: The Basis of Culture* (San Francisco: Ignatius Press, 2009).

189 **life of contemplation:** These are thoughts elaborated upon in this century by the theologian John Dunne. See, e.g., J. S. Dunne, *Love's Mind: An Essay on Contemplative Life* (Notre Dame, IN: University of Notre Dame Press, 1993).

190 **"holding ground":** This term is used by Philip Davis in *Reading and the Reader* (Oxford, UK: Oxford University Press, 2013).

190 "indifference toward meditative thinking": M. Heidegger, *Discourse on Thinking* (New York: Harper, 1966), 56.

190 "Digital media trains us": T. Wayne, "Our (Bare) Shelves, Our Selves," *New York Times*, Dec. 5, 2015.

191 "Readers know . . . in their bones": S. Wasserman, "The Fate of Books After the Age of Print," Truthdig, March 5, 2010, http://www.truthdig.com/arts_culture/item/steve_wasserman_on_the_fate_of_books_after_the_age_of_print_20100305/. Also in a different version in *Columbia Journalism Review*.

191 interview by Charlie Rose: Charlie Rose, interview, PBS, January 27, 2017.

192 "Where is the wisdom": T. S. Eliot, *Four Quartets* (New York: Harcourt, Brace & Company, 1943), 59.

193 "rhythm of time": I. Calvino, *Six Memos for the Next Millennium* (Cambridge, MA: Harvard University Press, 1988), 54.

194 Mrs. Ramsay: I would like to thank Andrew Piper in *Book Was There* for reminding me of the powerful example of reading in Virginia Woolf's novel *To the Lighthouse* (London: Hogarth Press, 1927).

194 Few historical individuals better illumine: I also include here Etty Hillesum, whose account of a concentration camp is extraordinary; see *An Interrupted Life: The Diaries and Letters of Etty Hillesum, 1941–1943*, introduction by J. G. Gaarlandt, trans. A. J. Pomerans (New York: Pantheon Books, 1984).

195 "Your prayers and kind thoughts": Quoted in E. Metaxas, *Bonhoeffer: Pastor, Martyr, Prophet, Spy* (Nashville: Thomas Nelson, 2010), 496.

195 "he always seemed to me": Ibid., 514, 528.

196 the Reader Organisation: I am reminded of the powerful contributions by prison volunteers such as the Reader Organisation in England that do the work of rehabilitating prisoners our society often doesn't, as well as helping the elderly and struggling students.

196 "a contentment and joy unparalleled": Personal interview, Providence, RI, 2014. See also B. Stiegler, Goldsmith Lectures, Lecture 1, 2013.

197 "The place of stillness": Quoted in L. Grossman, "Jonathan Franzen: Great American Novelist," *Time*, Aug. 12, 2010.

198 "I do believe": M. Robinson, *The Givenness of Things: Essays* (New York: Farrar, Straus and Giroux, 2015), 176, 187.

198 "hinge moment": Quoted in S. Wasserman's "The Fate of Books after the Age of Print," *Truthdig*, March 5, 2010.

199 "It would be catastrophic": M. Nussbaum, *Cultivating Humanity: A Classical Defense of Reform in Liberal Education* (Cambridge, MA: Harvard University Press, 1997), 300–01.

200 "If we look more closely": From *Letters and Papers from Prison*. First published in 1951, the English translation was published by

Touchstone Press in 1997. It is important to note that the first three words in the original title in German, *Widerstand und Ergebung: Briefe und Aufzeichnungen aus der Haft,* were left out in the English translation and describe the importance of taking a stand against the moral depravity in Nazism. I translate these words as Resistance and Resolution, though *Ergebung* also connotes what results from or develops out of taking an opposing stand of resistance.

200 **"The delicate game of democracy":** U. Eco and C. M. Martini, *Belief or Nonbelief? A Confrontation* (New York: Arcade Publishing, 2012), 71.

201 **a democracy succeeds only:** N. Strossen, *Hate: Why We Should Resist It with Free Speech, Not Censorship* (New York: Oxford University Press, 2018).

201 **The vacuum that occurs:** Throughout time demagogues and their loyalists have known the power of instilling fear, for those who fear make irrational choices about irrational fears. See the essay "Fear," *New York Review of Books,* Sept. 24, 2015, in which Marilynne Robinson wrote that fear can become an addiction. At the Nuremberg trials, Hermann Göring told the court that all one had to do to control any nation anytime was first to instill fear in the population and then to call anyone who disagreed a traitor. In our time, too many people call anyone who poses a threat to their views a liar. Whether the twentieth, the twenty-first, or any other century, when opposing ways of thinking are silenced, the "collective conscience" is gradually extinguished.

202 **combining of our highest intellectual:** See a different direction of work on empathy from the perspective of "reciprocal altruism" by Margaret Levi in "Reciprocal Altruism," Edge.org, Feb. 5, 2017, https://www.edge.org/response-detail/27170. She concludes, "Recognition of the significance of reciprocal altruism for the survival of a culture makes us aware of how dependent we are on each other. Sacrifices and giving, the stuff of altruism, are necessary ingredients for human cooperation, which itself is the basis of effective and thriving societies." See also her book with John Ahlquist *In the Interest of Others* (Princeton, NJ: Princeton University Press, 2013).

202 **"Wisdom, I conclude":** J. S. Dunne, *The House of Wisdom: A Pilgrimage* (New York: Harper & Row, 1985), 77. I see this passage as the modern complement to the line from Psalm 90: "Teach us to number our days that we may turn our hearts to wisdom."

202 **"There is always a 'feeling'":** C. Taylor, *The Language Animal: The Full Shape of the Human Linguistic Capacity* (Cambridge, MA: Belknap Press, 2016), 177. Please note that I have changed the translation in Taylor of the word *Laut*. Although it is rightly translated as "sound," I believe it is closer to Humboldt's intended meaning to use "speech."

203 **"possessing a language"**: Ibid.
203 **"the end of [the author's] wisdom"**: M. Proust, *On Reading*, ed. J. Autret, trans. W. Burford (New York: MacMillan, 1971; originally published 1906), 35.
203 **"Word-work is sublime"**: T. Morrison, "Nobel Lecture," Dec. 7, 1993, https://www.nobelprize.org/nobel_prizes/literature/laureates/1993/morrison-lecture.html.
204 **"act of resistance"**: Ulin, *The Lost Art of Reading*, 150.
205 **"endless forms most beautiful"**: From the beautiful quote in *On the Origin of Species* (1859): "There is grandeur in this view of life, with its several powers, having been originally breathed into a few forms or into one; and that, whilst this planet has gone cycling on according to the fixed law of gravity, from so simple a beginning endless forms most beautiful and most wonderful have been, and are being, evolved" (p. 490).

INDEX

Page numbers of illustrations appear in *italics*.

loss of internalized knowledge, 55–57, 89, 97, 117, 122–24, 201, 223n57
loss of meditative thinking, 190–92, 204
See also digital chain hypothesis
"digital wisdom," 177
Dirda, Michael, 34, 79, 150, 237n150
Doyle, Sir Arthur Conan, 60
dual-language learners, 158, 160, 170, 171, 177
Duckworth, Angela, 93–94, 229n93
Dunne, John S., 43, 69, 190, 192, 202
dyslexia, 6, 101, 155–56, 162, 164–65, 177, 180, 189, 218n29, 228n86, 238n155
Rapid Alternating Stimulus (RAS) test, 170–71

Eagleman, David, 16, 217n16, 219n34
East of Eden (Steinbeck), 40
Eco, Umberto, 200
Edelstein, Anne, 207, 208
Edmundson, Mark, 63, 74–75, 216n14, 224n63, 226n74
Einstein, Albert, 55
Eligible (Sittenfeld), 45
Eliot, George, 90, 92
Eliot, T. S., 192
Elman, Josh, 125
Else-Mitchell, Rose, 182
email, 72, 97, 98, 186
Emerson, Ralph Waldo, 56, 67, 225n67
empathy, 8, 31, 42–53, 61, 138, 162–64, 202, 221n43, 246n202
Enriquez, Juan, vii, 9, 169, 215n1
E-READ network, 126–27, 175–76
Evans, Barbara and Brad, 211, 212
Evolving Ourselves (Enriquez and Gullans), 9, 215n1

Facebook, 117, 122, 125
Fadiman, Anne, 101, 230n101
"Fear" (Robinson), 246n201
Ferrante, Elena, 189
festina lente, 193–94, 205
Fitzgerald, F. Scott, 41, 232–33n126
Fitzgerald, Penelope, 217n18
Flaubert, Gustave, 52
Fogassi, Leonardo, 59, 112, 231n112
Forster, E. M., 33, 42
Four Quartets (Eliot), 192
Francis, Pope, 53, 105, 178
Frank, Anne, 48, 164
Franzen, Jonathan, 197, 245n197
Freud, Sigmund, 63, 129
Friedlander, Elliott, 171
Frog and Toad (Lobel), 45
Frost, Joe, 137

Gaab, Nadine, 155, 238n154
Gabrieli, John, 155, 238n154
Galyean, Tinsley, 147
García Lorca, Federico, 103
Gardner, Howard, 109, 172, 241n172
Gates, Bill, 191
Geschwind, Norman, 219n31
Gilead (Robinson), 66, 224n66
Gioia, Dana, 73, 226n73
Gleason, Jean Berko, 139, 235n139
global literacy, 12, 147–48, 185, 216n12, 228n88
Goldenberg, Claude, 171, 241n171
Golinkoff, Roberta, 116, 232n116, 234n135
Goodman, Allegra, 103, 112
Goodnight Moon (Brown), 131
Goodwin, Doris Kearns, 189
Google, 85, 97, 123–24, 125, 232n125
Scholar, 97
Göring, Hermann, 246n201

ABOUT THE AUTHOR

MARYANNE WOLF is a scholar, mother, teacher, and advocate for children and literacy around the world. She is the director of the Center for Dyslexia, Diverse Learners, and Social Justice in the Graduate School of Education and Information Studies at UCLA; the Chapman University Presidential Fellow at Chapman University; and the former John DiBiaggio Professor of Citizenship and Public Service at Tufts University. She is the recipient of multiple research honors, including the Fulbright Fellowship and the highest awards by the International Dyslexia Association, the Australian Learning Disabilities Association, the Dyslexia Foundation, and the highest teaching awards from the Massachusetts and the American Psychological associations. She is the author of *Proust and the Squid: The Story and Science of the Reading Brain* (HarperCollins, 2007, translated in fourteen languages), *Dyslexia, Fluency, and the Brain* (editor, York, 2001), *Tales of Literacy for the 21st Century* (Oxford University Press, 2016), the *RAVE-O* reading curriculum, the *RAN/RAS* naming-speed tests (with Martha Denckla), and more than 160 scientific publications. One of the founding members of Curious Learning, she is involved in global literacy initiatives that help teach children to read, particularly in remote regions of the world and rural regions of the United States. She has given frequent lectures on literacy as a basic human right to the Pontifical Academy of Sciences at the Vatican and serves as an external advisor to multiple boards, including the Canadian Children's Literacy Foundation and the International Monetary Fund. She has two sons and lives in Los Angeles.

ALSO BY MARYANNE WOLF

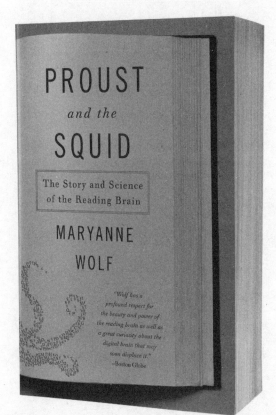

PROUST AND THE SQUID
THE STORY AND SCIENCE OF THE READING BRAIN

"[Maryanne Wolf] displays extraordinary passion and perceptiveness concerning the reading brain, its miraculous achievements and tragic dysfunctions." —*BookForum*

Lively, erudite, and rich with examples, *Proust and the Squid* asserts that the brain that examined the tiny clay tablets of the Sumerians was a very different brain from the one that is immersed in today's technology-driven literacy. The potential transformations in this changed reading brain, Wolf argues, have profound implications for every child and for the intellectual development of our species.

ALSO AVAILABLE IN PAPERBACK, EBOOK, AND DIGITAL AUDIO